T0155504

Optimization of Process Flowsheets through Metaheuristic Techniques

José María Ponce-Ortega
Luis Germán Hernández-Pérez

Optimization of Process Flowsheets through Metaheuristic Techniques

José María Ponce-Ortega
Universidad Michoacana
de San Nicolás de Hidalgo
Morelia, Michoacán, Mexico

Luis Germán Hernández-Pérez
Universidad Michoacana
de San Nicolás de Hidalgo
Morelia, Michoacán, Mexico

ISBN 978-3-030-06284-2 ISBN 978-3-319-91722-1 (eBook)
https://doi.org/10.1007/978-3-319-91722-1

Printed on acid-free paper

This Springer imprint is published by the registered company Springer International Publishing AG part of Springer Nature.
The registered company address is: Gewerbestrasse 11, 6330 Cham, Switzerland

Preface

This book presents a general framework to implement a link between process simulators and optimization through metaheuristic techniques. The book describes step-by-step the methodology to implement this link for different process simulators and with different metaheuristic methods.

The aim of this book is to provide the readers the needed knowledge to implement optimizations of process flowsheets through links between process simulators and metaheuristic approaches. This way, basic knowledge about simulation through process simulators is needed. To implement this link between process simulation and metaheuristic techniques, the approach is divided into three fundamental sections: process simulation, metaheuristic algorithm, and implementation of the link between process simulation and optimization, which are described in the following chapters.

Chapter 1 presents some basic concepts needed. Chapter 2 presents an introduction about the general concepts that are involved in the process simulation and the main commercial software currently available to efficiently carry out this function. Chapter 2 also presents the basics about the management to manipulate simulations of chemical and industrial processes.

Chapter 3 presents an introduction about metaheuristic optimization methods, which can be then included in the link to process simulators and optimization. Chapter 4 explains how to implement the link between the process simulators and optimization programs containing metaheuristic techniques, which correspond to the optimization of the flowsheet of the simulation of the process to be optimized. Chapter 4 also presents a detailed explanation of the presented methodology to implement the link between process simulators and optimization, which corresponds to the linking of programs. This part of the book is the main contribution of the proposed methodology. For its better understanding, the steps of the proposed methodology are first explained. Then, the needed code is provided to implement the appropriate link between simulation software and stochastic algorithms. For this purpose, the sequence to be followed is mentioned step by step, indicating how to call the needed variables.

Chapter 5 shows the evaluation performance of the different software considered for implementing the link for the optimization of process flowsheets through process simulators and metaheuristic techniques.

Chapters 6 and 7 show two case studies to present the application of the proposed methodology. Chapter 6 shows the optimization of an industrial process (steam power plant in Aspen Plus®). In the same way, Chap. 7 shows the optimization of another industrial process (biodiesel in SuperPro Designer®).

This book also includes some tutorial videos that show, step by step, the proposed methodology to implement a link between process simulators and optimization through metaheuristic optimization approaches. These videos are prepared to show the implementation of the proposed approach for different process simulators and with different alternatives to implement the metaheuristic approach.

Morelia, Michoacán, Mexico

José María Ponce-Ortega
Luis Germán Hernández-Pérez

Abstract

This book presents a multi-objective optimization framework for optimizing chemical processes. The proposed framework implements a link between process simulators and metaheuristic techniques. The proposed approach is general, and there can be used any process simulator and any metaheuristic technique. This book shows how to implement links between different process simulators such as Aspen Plus®, HYSYS®, SuperPro Designer®, and others, linked to metaheuristic techniques implemented in Matlab®, Excel®, C++, or other programs. This way, the proposed framework allows optimizing any process flowsheet implemented in the process simulator and using the metaheuristic technique, and this way the numerical complications through the optimization process can be eliminated. Furthermore, the proposed framework allows using the thermodynamic, design, and constitutive equations implemented in the process simulator to implement any process.

Keywords: Optimal design, Metaheuristic optimization, Multi-objective optimization, Process simulators, Simulation

Contents

List of Figures

List of Table

Chapter 1
Introduction

The fundamental concepts used in this book are described below. To implement the link between any process simulator and metaheuristic techniques, the methodology has been divided in three parts: simulation, optimization, and link software; and the involved concepts are described as follows.

1.1 Process Simulation

A mathematical model of a chemical process is a simplified representation of the physicochemical behavior of a real process, which is used to predict values of output variables for given input variables and process design (including operating) variables. A model can be used for what-if studies and process troubleshooting, and it has many applications for process optimization, process control, and operator training. Models are often difficult to solve analytically, and so they are mostly solved numerically (Sharma and Rangaiah 2016).

Process simulation allows predicting the behavior of a process by using basic engineering relationships, such as mass and energy balances, and phase and chemical equilibrium. Given reliable thermodynamic data, realistic operating conditions, and rigorous equipment models, one can simulate the plant behavior. Process simulation enables to run many cases, conduct "what-if" analyses, and perform sensitivity studies and optimization runs. With simulation, one can design better plants and increase profitability in existing plants. Process simulation is useful throughout the entire life cycle of a process, from research and development through process design to production (AspenTech 2015).

Modeling refers to all the steps involved in developing and validating a model for the process, whereas simulation refers to the use of the developed model for studying the process behavior/response for one or more sets of input and design variables. In general, modeling and simulation are used to optimize the process

J. M. Ponce-Ortega, L. G. Hernández-Pérez, *Optimization of Process Flowsheets through Metaheuristic Techniques*,
https://doi.org/10.1007/978-3-319-91722-1_1

operation and design. Optimization improves the performance of a process by changing the operating conditions such as temperature, pressure, and flow rate of process streams but without changing the size of any equipment or process flowsheet. Process retrofitting and revamping refer to redesign a plant for specific objective(s), such as increasing throughput, decreasing energy consumption, and revising product quality. This is achieved by changes in existing equipment and/or addition of new equipment (leading to a new process configuration) besides changes in operating conditions (Sharma and Rangaiah 2016).

1.2 Searching Methods

Optimization is the act of making something as good as possible (Cambridge dictionary). The word optimum means "the best." Optimization consists in finding the optimum point in which the best values are found for certain variables and in which they achieve some specific objectives. There are different search methods to achieve optimization, which will be analyzed in depth in the corresponding chapter.

The decision variables correspond to those that have been previously determined and will be manipulated in order to find the optimal point. These variables must operate within a range in which they offer feasible results for the objectives being sought; this is the operating range. Also, the decision variables can be subject to certain constraints that the user must know for the studied processes; these constraints help to limit the search range and make the optimization more efficient, reducing the computation time.

The objective function is an equation in which is reflected the performance of the process that is being optimized; it is achieving its maximum or its minimum value by manipulating the variables of dissolution and considering the established search restrictions.

1.2.1 Classification of Search Methods

General search and optimization techniques are classified into three categories: enumerative, deterministic, and stochastic (Coello-Coello et al. 2002). Figure 1.1 shows common examples of each type. Some authors classify calculus-based methods in indirect and direct, and classify evolutionary computation in evolution strategies, evolutionary programming, genetic algorithms, and genetic programing (Devillers 1996).

To overcome some drawbacks associated to the deterministic optimization approaches, metaheuristic optimization approaches have been reported (Wang and Tang 2013; Guo et al. 2014; Ouyang et al. 2015; Wong et al. 2016). The metaheuristic optimization approaches mimic some evolution processes and are based on

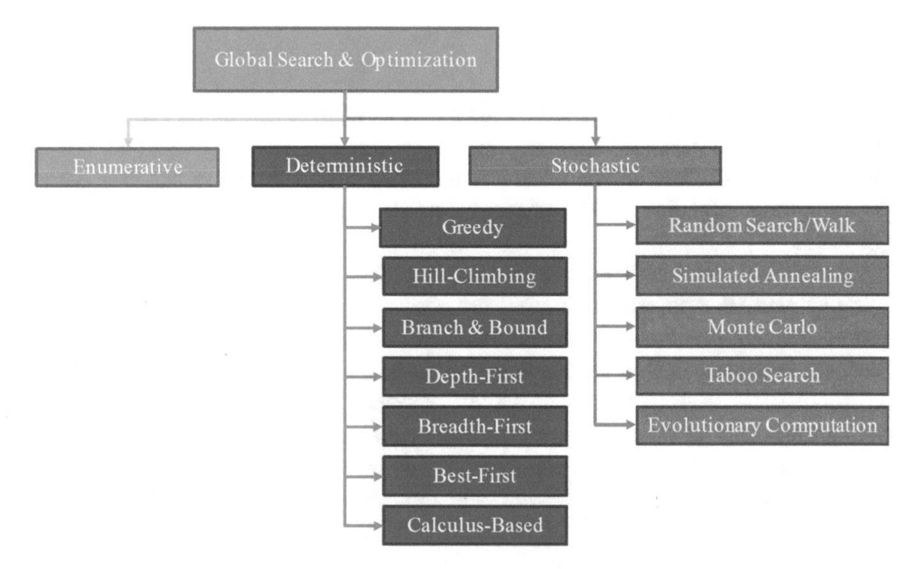

Fig. 1.1 Global optimization approaches and different classes of search methods (Coello-Coello et al. 2002; Devillers 1996)

repeatedly simulating a given process to assign a fitness function for given values of the degrees of freedom (Sharma and Rangaiah 2016). This way, the possibility to get trapped prematurely in a suboptimal solution is avoided (Devillers 1996), the complication associated to the form of the optimization model is avoided (Sharma and Rangaiah 2014), and mainly because these are based on simulation of the process, these can be easily linked to the process simulators to optimize different flowsheets.

1.2.2 Deterministic Algorithms

Deterministic methods have been successfully used in solving a wide variety of problems. However, these methods are inefficient for solving non-convex and non-linear problems. For the implementation of these optimization methods, it is necessary to implement all the equations that describe the behavior of the process by the formulation of a mathematical model.

Deterministic methods are often ineffective when applied to NP-complete or other high-dimensional problems because they are limited by their requirements associated to the problem domain, knowledge (heuristics), and the search space, which can be exceptionally large. Because many real-world scientific and engineering multi-objective problems (MOPs) exhibit one or more of the abovementioned characteristics, stochastic searches have been developed as alternative approaches for solving these irregular problems (Coello-Coello et al. 2002).

1.3 Interaction Between Programs

When the word link is mentioned in this book, it refers to the relationship between programs for the purpose of controlling or sending and receiving data obtained in different software. The link between programs can be established through the use of internal tools of some of the programs or through the instructions of a third program.

Visual Basic for Applications (VBA) is the language of the Microsoft (MS) operating system (Windows) that is used for program applications. Many of the programs and add-ons that are used in Windows are developed in this language, so there are common elements that can be manipulated through this platform.

Microsoft COM (Component Object Model) technology in the Microsoft Windows-family of operating systems that enables software components to communicate (microsoft.com), for this reason, this technology is used to achieve the link between the simulation software and the program in which the optimization algorithm is based. The details of the use of the COM technology will be described in the corresponding chapter.

1.4 Nomenclature

ACM	Aspen Custom Modeler
COM	Component Object Module
MOP	Multi-objective problems
MS	Microsoft®
VBA	Visual Basic for Application

Chapter 2
Process Simulators

Process simulator is a computer program that allows modeling different processes depending on the area of study for which it was designed; this way, there are process simulators for industrial, chemical, and biochemical processes. A process simulation software is the best way to perform the simulation of industrial processes; this is due to the large number of equations and numerical methods that are needed to use for proper representation and prediction of behavior in reality.

In addition, the process simulators usually are programmed for using in the operating system of a computer, so it is advisable to verify the compatibility of the software that will be used with the equipment where it is going to work. Currently, there are several process simulators that are distributed commercially and which already have modeling equations for certain equipment and numerical methods programmed for the solution of specific thermodynamic equations. Another important aspect of commercial simulation software is that they have a simple database with components usually used in the chemical and process industry, as well as their physicochemical and thermodynamic properties.

Process simulators have been widely used for analyzing chemical processes (Morgan et al. 2017; Gómez-Ríos et al. 2017; Pauls et al. 2016); these offer a tremendous advantage associated with the implemented thermodynamic correlations as well as the powerful numerical methods for solving the mass and energy balances together with design and constitutive relationships (Hauck et al. 2017). Process simulators allow analyzing the process flowsheets for zero degrees of freedom. Some optimization approaches have been implemented in process simulators; however, important drawbacks have been identified associated with the number of degrees of freedom, the use of explicit constraints, as well as the number of objective functions. Furthermore, the implemented optimization techniques can be prematurely trapped in suboptimal solutions, or even no feasible solutions can be obtained (Coello-Coello et al. 2002) because usually deterministic optimization techniques have been implemented (Ponce-Ortega et al. 2012).

J. M. Ponce-Ortega, L. G. Hernández-Pérez, *Optimization of Process Flowsheets through Metaheuristic Techniques*,
https://doi.org/10.1007/978-3-319-91722-1_2

Nowadays, several process simulators, such as Aspen Plus® and Aspen HYSYS®, are commercially available for simulating complete chemical processes, where common process units and a property database for numerous chemicals are available (Sharma and Rangaiah 2016). Most of chemical and biochemical process simulators are not equipped with adequate optimization tools. However, in very few simulators (e.g., Aspen Plus®), there are some optimization tools, but the formulation of optimization problems and available solution techniques is not good enough (Woinaroschy 2009). For example, the number of degrees of freedom is limited, only deterministic techniques can be implemented, it is complicated to manipulate explicit constraints for not manipulated variables, and only one objective can be considered.

For the proposed multi-objective optimization framework, the first step consists in implementing the flowsheet in the process simulator. The input variables and the operating conditions for the included equipment must be specified. Also, the thermodynamic method and the mathematical solution technique with the maximum number of iterations are necessary to be declared. All these values are for the first simulation process, and it must be run without any error or warning. It is recommended to validate the response variables for checking the values of the results after implementing the optimization approach.

2.1 Aspen Plus®

Aspen Plus® is the market-leading chemical process simulation software used by the bulk, specialty, and biochemical industries for the design and operation (aspentech.com). The main advantages of this simulator consist of a large database of specific chemical compounds and unit operations.

However, models for less common and/or new process units are not readily available in the simulators, but they may be available in the literature or can be developed from first principles. A mathematical model for a new process unit can be implemented in Aspen Custom Modeler (ACM), and then it can be exported to Aspen Plus® or Aspen HYSYS® for simulating processes, having a new process unit besides common process units such as heat exchangers, compressors, reactors, and columns (Sharma and Rangaiah 2016).

For the correct functioning of these simulations, it is necessary to feed the program with values that are within a suitable range, the previous one in order to avoid errors in the equipment so that indeterminacies arise due to the thermodynamic behavior of the substances used and the interconnections of the equipment must be correctly indicated.

Aspen Plus® is a process simulation program that can also be used for many types of thermodynamic calculations or to retrieve and/or correlate thermodynamic and transport data (Sandler 2015). With the purpose of obtaining a better understanding of the use of this software for process simulation, we will present some fundamental aspects for its use. However, it is worth noting that if the reader requires further information about the use of this specific software, it is better to consult the user guide that the program developers offer on the official website. As the Aspen Plus® V8.8 Help indicates, one can translate any process into an Aspen Plus® process simulation model by performing the following steps:

1. Specify the chemical components used in the process. You can take these components from the Aspen Plus® databanks, or you can define them.
2. Specify the thermodynamic models to represent the physical properties of the components and mixtures in the process. These models are included in the Aspen Plus® software.
3. Define the process flowsheet, which includes:
   (a) Define the unit operations in the process.
   (b) Define the process streams that flow to and from the unit operations.
   (c) Select models from the Aspen Plus® Model Library to describe each unit operation and place them on the process flowsheet.
   (d) Place labeled streams on the process flowsheet and connect them to the unit operation models.
4. Specify the component flow rates and the thermodynamic conditions (e.g., temperature and pressure) of feed streams.
5. Specify the operating conditions for the unit operation models.

With Aspen Plus®, one can interactively change specifications, such as flowsheet configuration, operating conditions, and feed compositions, to run new cases and analyze process alternatives. In addition to process simulation, Aspen Plus® allows to perform a wide range of other tasks such as estimating and regressing physical properties, generating custom graphical and tabular output results, fitting plant data to simulation models, optimizing processes, and interfacing results to spreadsheets (AspenTech 2015).

The user interface of Aspen Plus® is very intuitive and easy to use, due to the remarkable efforts that the developers of this software have implemented to make the use friendlier. This important aspect can be noticed with the improvements that the new version has with respect to the previous one.

To start, open the Aspen Plus V8.x, which you may have to locate depending on the setup of your computer. (It may be on your desktop or you may have to follow the path All Programs > Aspen Tech > Process Modeling V8.x > Aspen Plus > Aspen Plus V8.x.)

When you open Aspen Plus V8.2 or higher version, you will briefly see the Aspen logo of Fig. 2.1. There is then a slight delay while the program connects to the server, and then the Start Using Aspen Plus window with resent simulations appears.

To proceed, click on New, which brings up the window shown in Fig. 2.2 for all versions of Aspen Plus V8.0 or higher.

Click on Blank Simulation and then Create. This will bring up Fig. 2.3.

On the lower-left-hand corner of this window, there are three choices. The first one, which Aspen Plus opens, is Properties; the drop-down menu under Components > Specifications is used to specify the component or components for

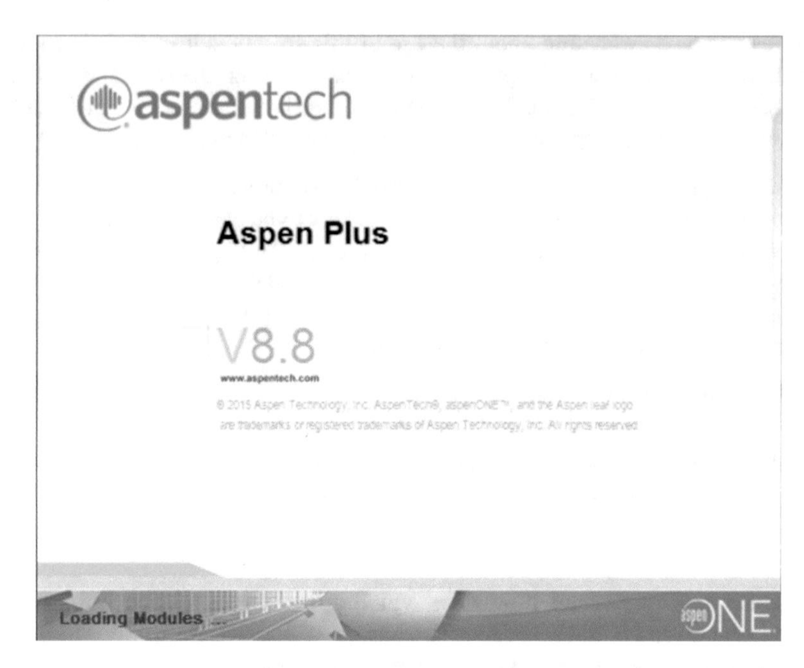

Fig. 2.1 Aspen Plus V8.8 Start-up

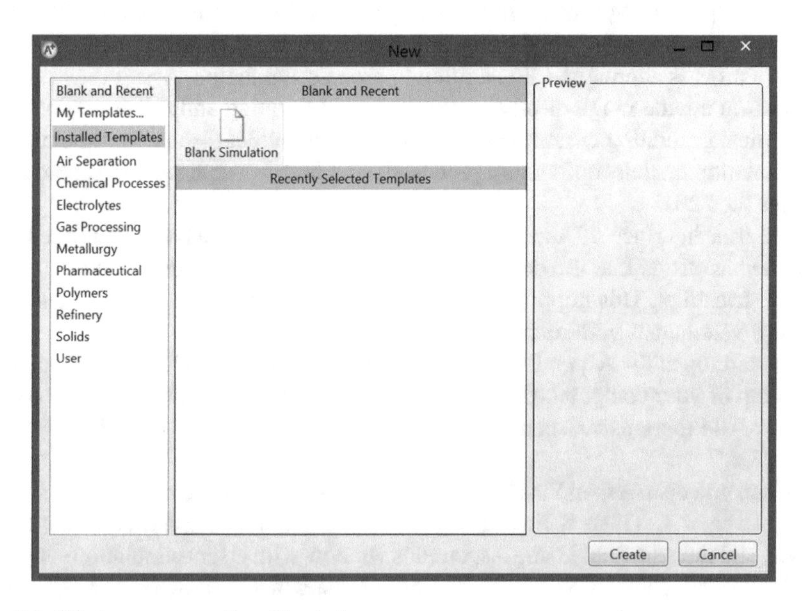

Fig. 2.2 Window to open a New Simulation

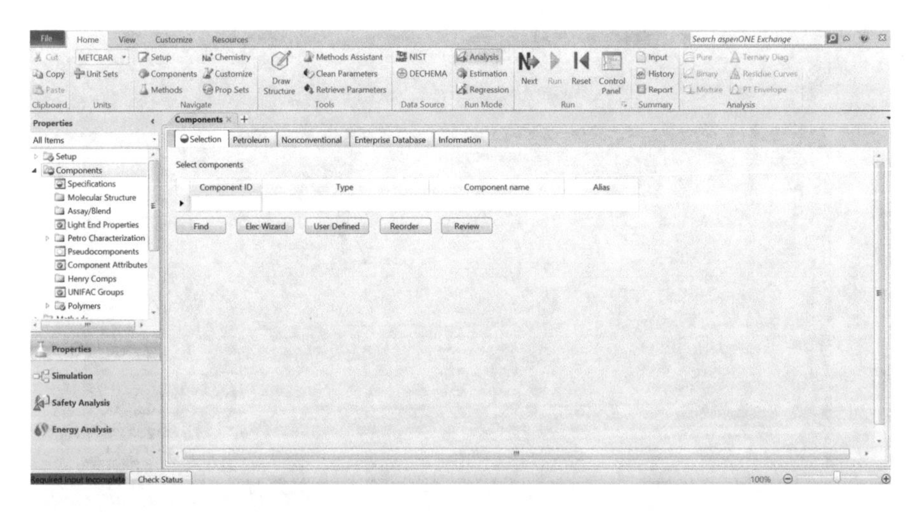

Fig. 2.3 Properties of Aspen Plus

the calculation, and the drop-down menu under Methods is used to specify the thermodynamic models and parameters that will be used in the calculation. The second general area is Simulation that will take you to a flowsheet window, to be discussed later, and the third one is the Energy Analysis that is used to implement energetic analysis and integration. The default is to start with Properties.

For example, we will proceed to entering the component water. There are two ways to enter component names. The simples and most reliable way to ensure that you will get the correct component and its properties from the Aspen Plus® database is to click on the Find box that brings up the Find Compounds window and to enter the component name by typing in water and then clicking on Find Now, which produces the window shown in Fig. 2.4.

A long list of compounds, shown in Fig. 2.4, is generated because the default Contains was used in the Find Compounds window; as a result, every compound in the database that contains water either in its compound name or in its alternate name appears in the list. The compound we are interested will be first on this list, but that will not always be the case. Therefore, a better way to proceed in the Find Compounds window is to click Equals instead of the default Contains and then click Find Now, which produces, instead of a list, only water (Fig. 2.5).

Click on WATER and the Add selected compounds, and for this example, click on Close. You will then see Fig. 2.6.

Another alternative is to type in all or part of the name directly in the Components-Specification window and see whether Aspen Plus® finds the correct name. Notice that water has now been added to the Select components list and that both components and Specifications now have check marks indicating that sufficient information has been provided to proceed to the next step. However, this may not be sufficient information for the problem of interest to the user. If the problem to be solved involves a mixture, one or more additional components may be added following the

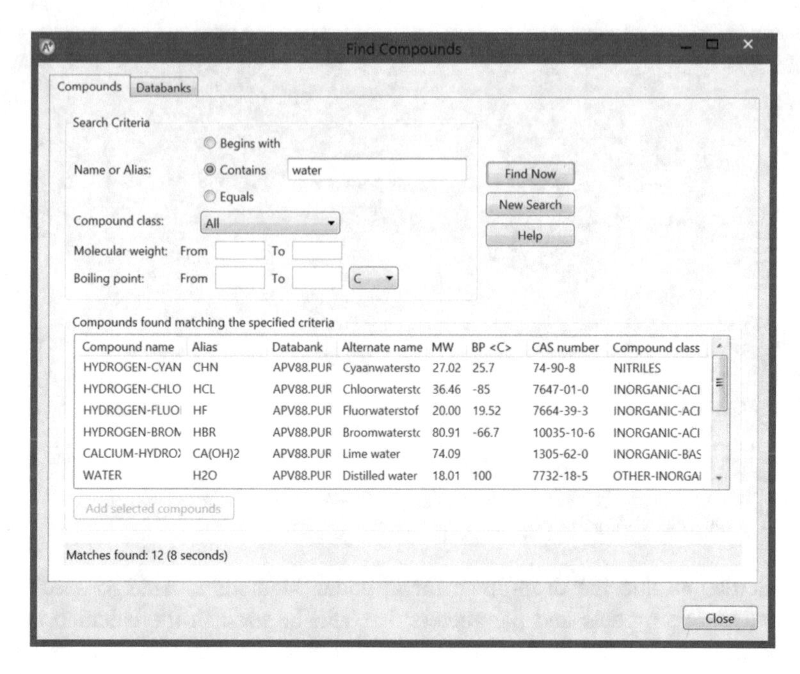

Fig. 2.4 Find Component Window (Contains option)

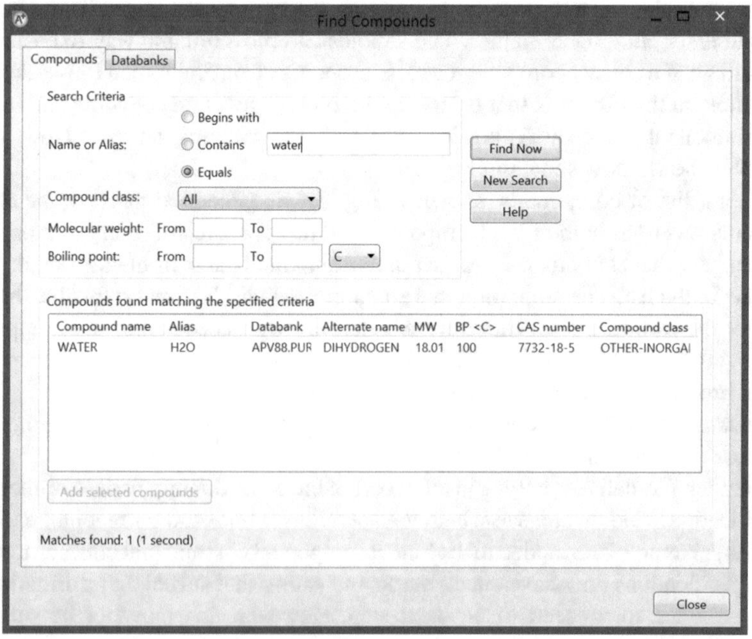

Fig. 2.5 Find Component Window (Equals option)

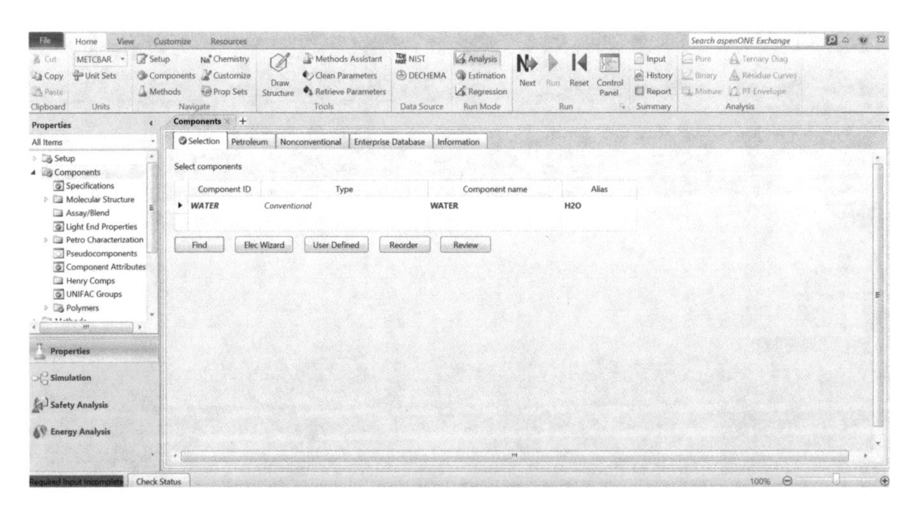

Fig. 2.6 Component window

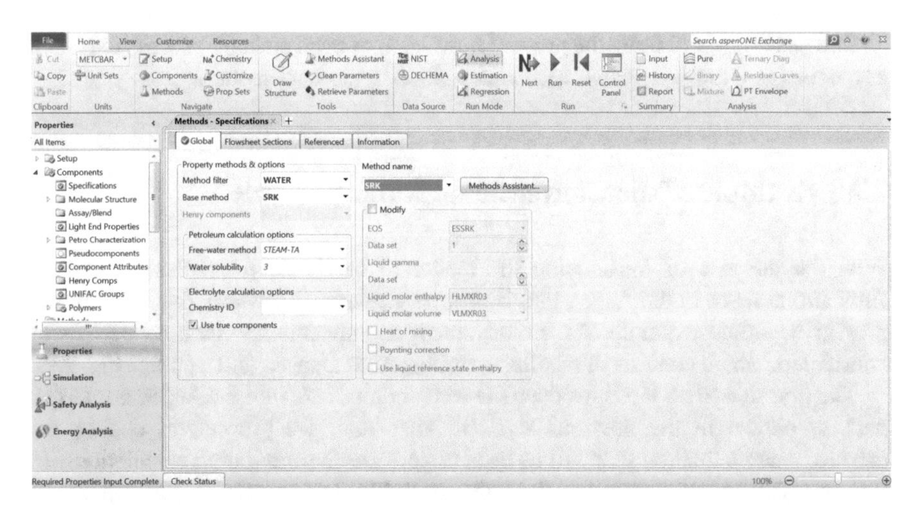

Fig. 2.7 Methods window

procedure described above except that the Close button in the Find Compounds window is used only after all the components have been added.

The next step is to go to Methods by clicking on it. The window shown in Fig. 2.7 appears, and here a number of thermodynamic models can be used through any other equation of state for which parameters are available can be used. Notice that if you need help in choosing a thermodynamic model, you can click on Methods Assistant for help. After accepting the equation by clicking Enter, Methods on the left-hand side of Fig. 2.7 will also have a check.

Checking on Simulation brings up the Main Flowsheet window of Fig. 2.8 together with Model palette at the bottom of the window.

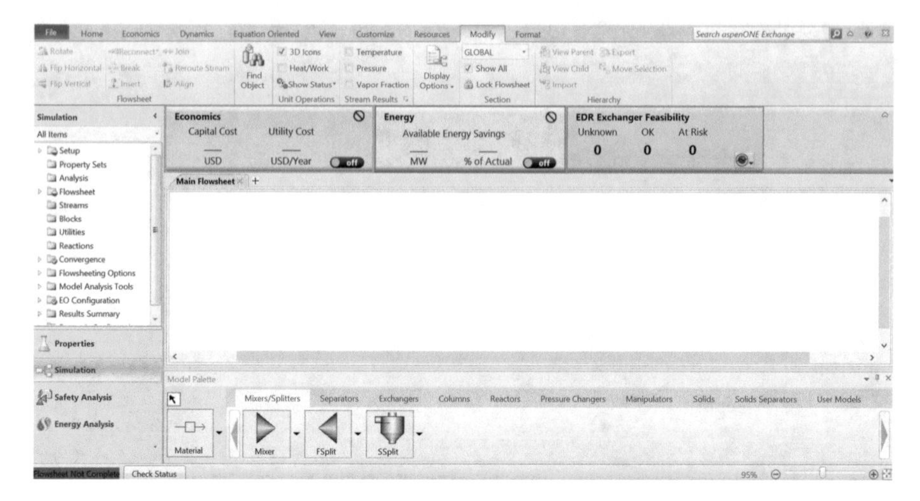

Fig. 2.8 Simulation section of Aspen Plus

The next step is to draw the process flowsheet, or even a single process unit such as a boiler or a turbine, which will be described in the next section.

2.2 Example of the Conventional Rankine Cycle

With the purpose of introducing the reader to the basic principles about how simulations work in the Aspen Plus® process simulator software, this section will present a simple example. As a case study, a conventional Rankine cycle was considered, which consists of a boiler, a turbine, a condenser, and a pump (Fig. 2.9).

The first step to do is to introduce the water component into the Aspen properties part, as shown in the previous section. After that, we proceed to choose the thermodynamic method that will be used to perform the simulation calculations (in this case STEAMNBS with a method filter of WATER). Once this is done, we select the Aspen Simulation part, where we proceed to build the process flowsheet selecting each equipment that conforms the process of the conventional Rankine cycle. To select each equipment, go to the model palette (in case this is not visible, go to the display tab in the show section and select model palette) (Fig. 2.10).

The first unit for the flowsheet process of a conventional Rankine cycle is the boiler. You have to go to the model palette, in the heat exchanger tab, and look for the boiler symbol. To select it, you must click on the corresponding symbol and click again in any part of the work area of the Aspen simulation part (as Fig. 2.11 shows). A small symbol will appear as the one with a default name, which will be "B1"; to change it just click on the symbol and with right click select the "Rename Block" option. Now we can assign a more appropriate name to better identify it; in this case, it can be "BOILER."

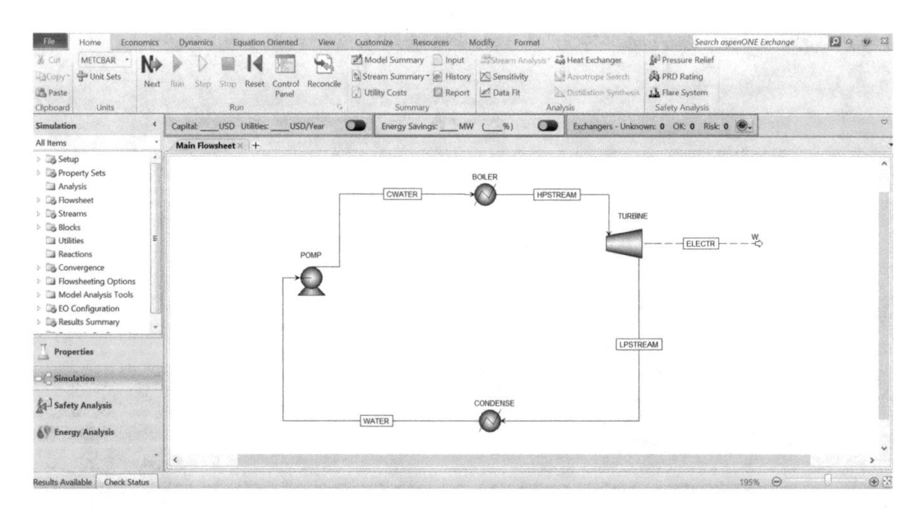

Fig. 2.9 Rankine cycle flowsheet in Aspen Plus®

Fig. 2.10 Location of the model palette

In the same way, we proceed to select and rename the rest and the necessary equipment for the flowsheet of the process of a conventional Rankine cycle (Fig. 2.12).

The next step is to complete the processes flowsheet of the conventional Rankine cycle joining the blocks using material streams for that. To do this, select the Material option from the model palette with a click. You will notice that in the blocks of the diagram, there will appear small red and blue arrows; the first common meaning is obligatory to complete the process flowsheet, while the blue ones are optional (Fig. 2.13).

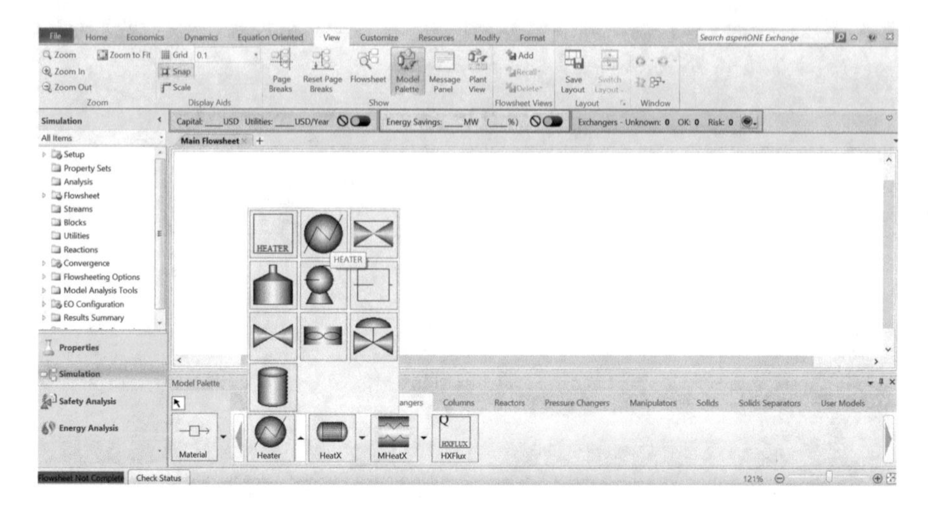

Fig. 2.11 Location of the boiler in the model palette

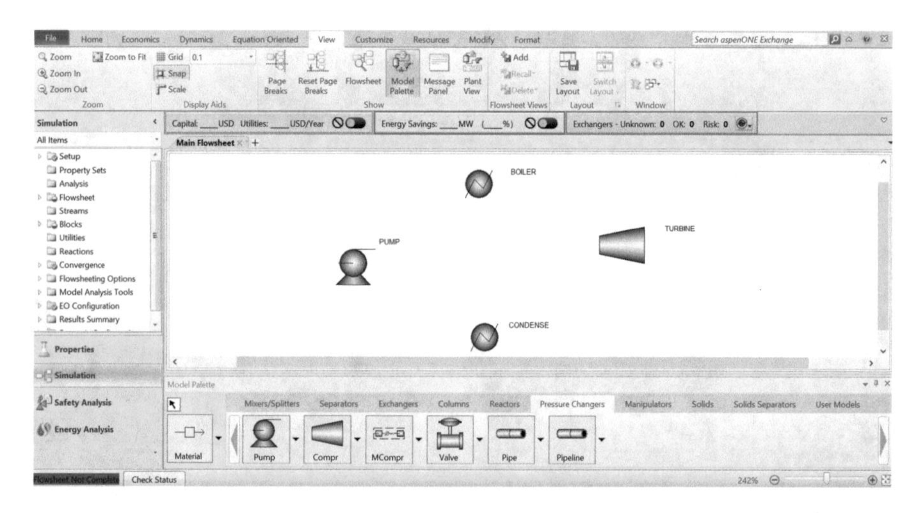

Fig. 2.12 Necessary blocks for the conventional Rankine cycle

Connect the blocks using the material streams. This is done by clicking and holding on an arrow, moving the pointer to the desired location, and releasing it. Rename the streams to obtain the process flowsheet of the conventional Rankine cycle shown in Fig. 2.9.

To proceed, the user must now provide the specifications for the process, which includes the inlet stream component(s) and conditions, pressures and temperatures needed, and the type of each process unit or block (e.g., an isentropic turbine). Click on Streams > WATER and complete with a temperature of 98 °C, a pressure of 1 atm, a total flow rate of 20,000 kg/h, and a molar fraction of 1 in water compound, as shown in Fig. 2.14. This is the only stream that needs to be specified;

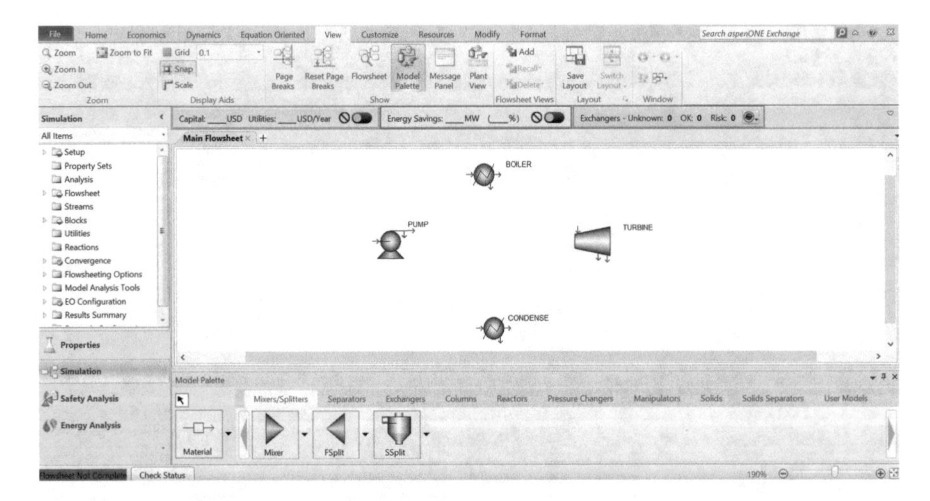

Fig. 2.13 Appearance of the blocks when the stream option is selected

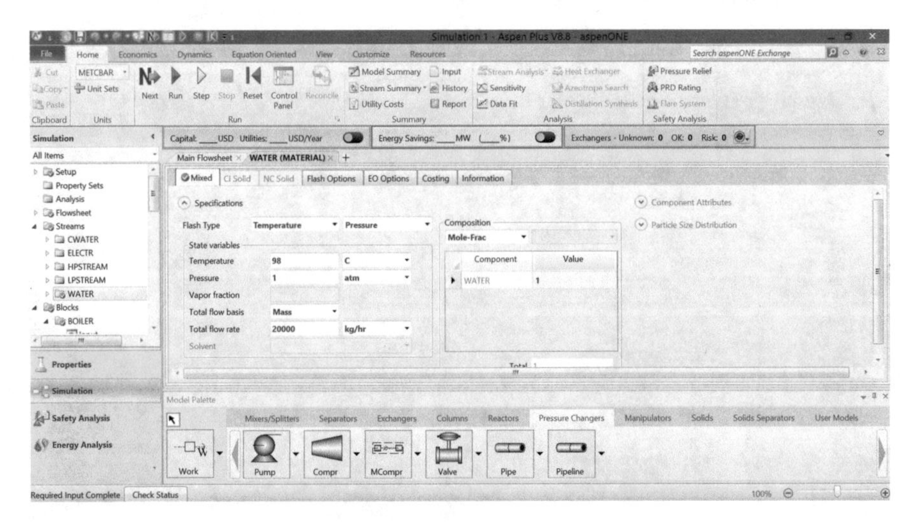

Fig. 2.14 Specification of the WATER stream values

the properties of all the other streams will be computed in the simulation calculation once the actions of the Blocks are chosen.

We then move on to the specifications for the Blocks (process units). First, for the BOILER, we specify a temperature of 460 °C and a pressure of 40 atm (Fig. 2.15).

Next in the block CONDENSE (the condenser), which is a heat exchanger, there are specified the operating conditions, which are a temperature of 98 °C and a pressure of 1 atm (Fig. 2.16).

The pump, PUMP, is operated to discharge a pressure of 40 atm (Fig. 2.17). Other entries are the unchanged defaults.

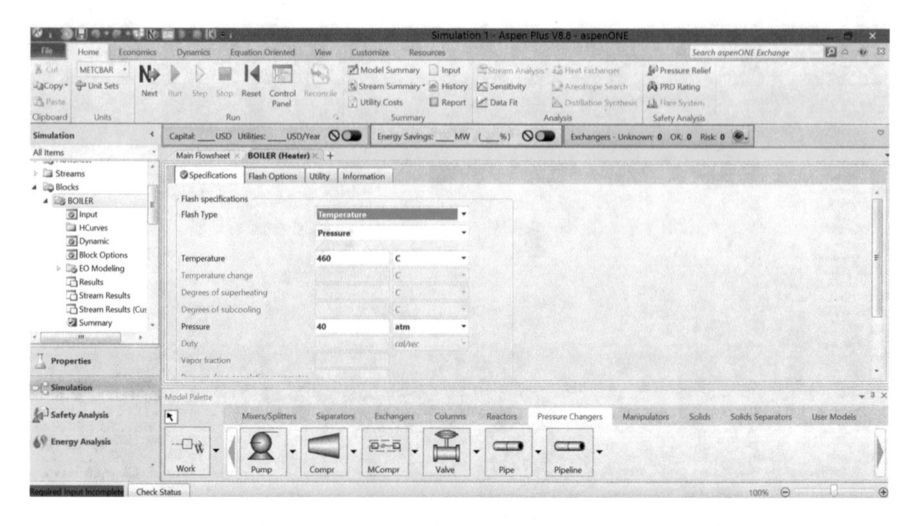

Fig. 2.15 Specification of the BOILER block values

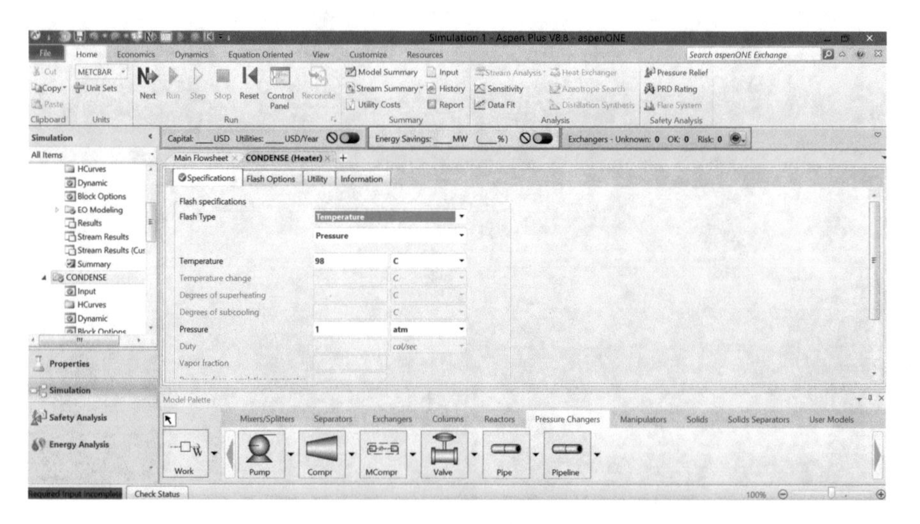

Fig. 2.16 Specification of the CONDENSE block values

Finally, the specifications for the turbine, TURBINE, are that it is operated as an isentropic turbine and it discharges at a pressure of 1 atm (Fig. 2.18).

Now all necessary boxes are checked, and in the lower-right-hand corner of the window, there is the message "Required Input Complete." We are ready to run the simulation. There are five ways to start the simulation. The first way is to press the F5 key (not function F5, just the F5 key). The second and third ways are to press one of the two forward arrow keys on the Main Toolbar and click on the forward arrow above Run (which will gray if not all the information for the simulation has been entered, but turn dark blue when all necessary data have been entered). The final two

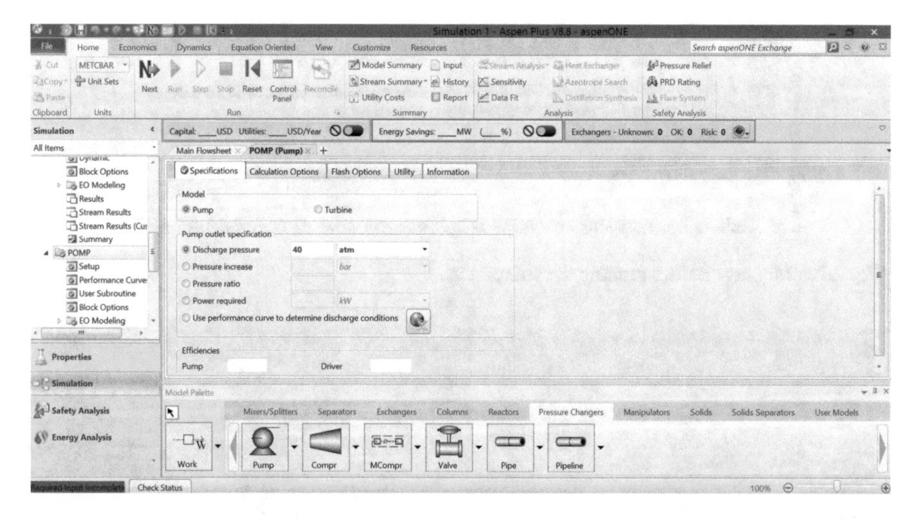

Fig. 2.17 Specification of the PUMP block values

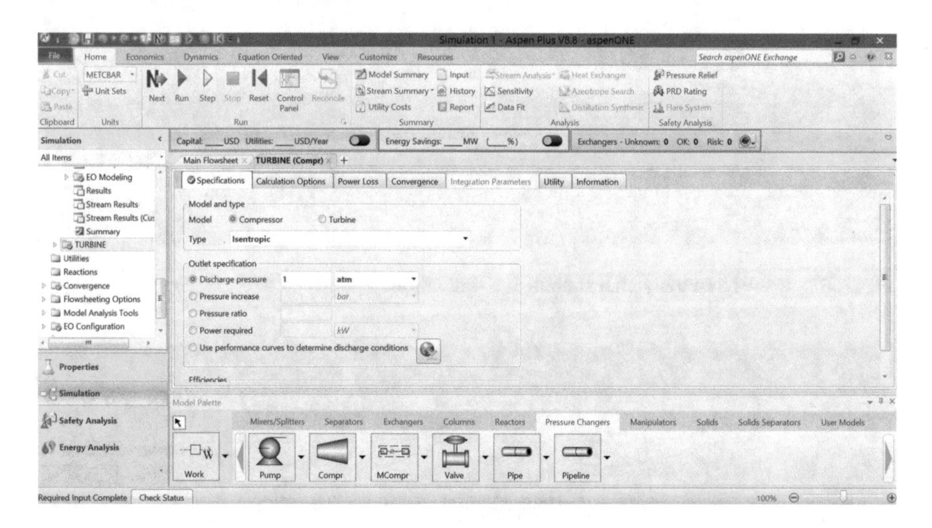

Fig. 2.18 Specification of the TURBINE block values

ways are to click on one of the Aspen Plus Next keys on the Main Toolbar that will bring up the message of Fig. 2.19. Clicking OK will then run the simulation.

After running the simulation, you can check the results clicking on Results Summary (Fig. 2.20).

Then, going to Streams, it brings up a window (Fig. 2.21) containing the table of stream results.

As you can see, Fig. 2.21 shows the Material Streams. If you want to see the work generated in the turbine, you can see it clicking on the Work tab and it brings up a window with the value (Fig. 2.22).

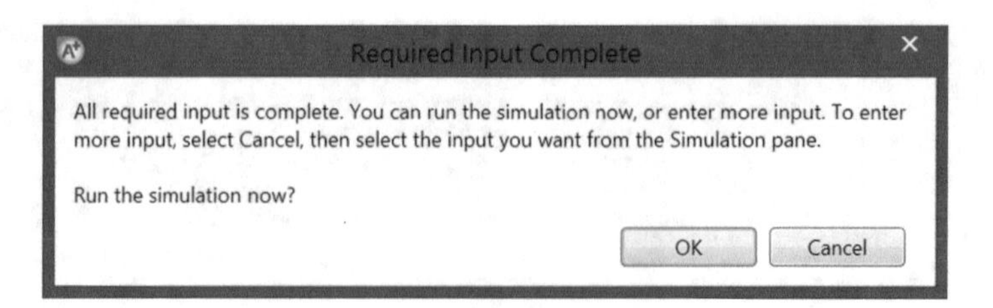

Fig. 2.19 Message before running the simulation

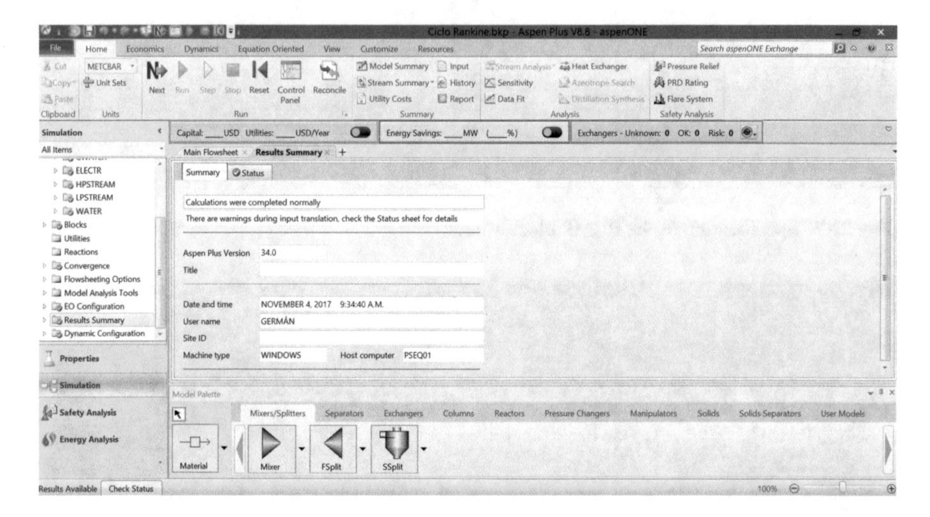

Fig. 2.20 Results Summary after running the simulation

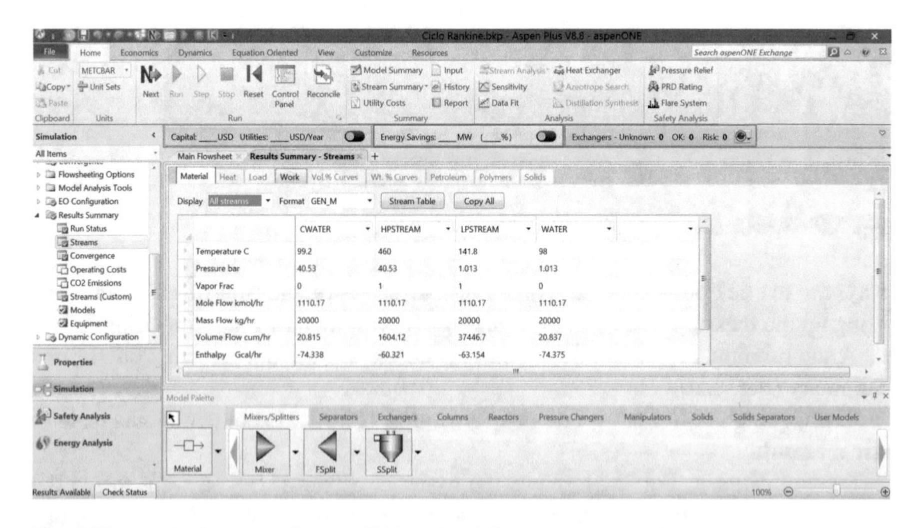

Fig. 2.21 Results Summary Streams Table (Material)

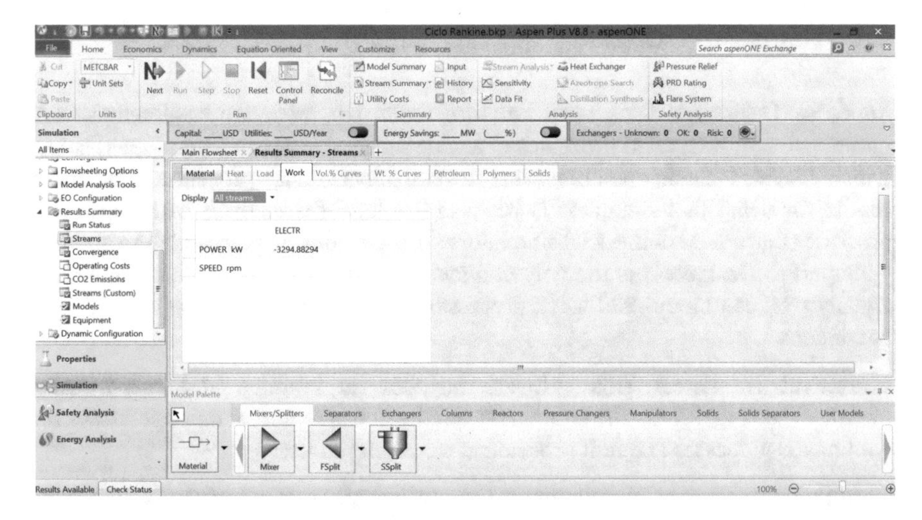

Fig. 2.22 Results Summary Streams Table (Work)

As you can see, Aspen Plus® is a process simulator software with a user interface very easy and intuitive to manipulate. If you need to deep in the use of Aspen Plus, it is recommended to consult a specialized text as mentioned in the bibliography of this book.

2.3 Aspen HYSYS®

Aspen HYSYS® is a process simulation software focused on oil and gas, refining, and engineering processes (aspentech.com). With an extensive array of unit operations, specialized work environments, and a robust solver, modeling in Aspen HYSYS V8 enables the user to:

- Improve equipment design and performance
- Monitor safety and operational issues in the plant
- Analyze processing capacity and operating conditions
- Identify energy savings opportunities and reduce greenhouse gas emissions
- Perform economic evaluation to obtain savings in the process design

Aspen HYSYS V8 builds upon the legacy modeling environment, adding increased value with integrated products and an improved user experience. The ease of use and flexibility of model calculations have been preserved, while new capabilities have also been added.

2.4 SuperPro Designer®

SuperPro Designer® is a software that facilitates modeling, evaluation, and optimization of integrated processes in a wide range of industries (intelligent.com), which includes batch operations and several biochemical processes. The main reason for using this simulator is because it allows the analysis of biochemical processes that other commercial simulators do not include in their modeling options.

SuperPro Designer® is the only commercial process simulator that can handle equally well continuous and batch processes as well as combinations of batch and continuous.

Graphical Interface: This software includes an intuitive and user-friendly interface (see Fig. 2.23). The equipment-looking icons represent unit operations for continuous processes and unit procedures for batch processes.

In this environment, developing a process flowsheet or modifying values is as easy as point and click. The interface is very similar to other MS Windows applications, making its features very intuitive.

Unit Procedures: A unit procedure is a set of operations that take place sequentially in a piece of equipment. For instance, the P-1 vessel unit procedure (see Fig. 2.24) includes the following operations: Charge Solvent, Charge Reactant A, Charge Reactant B, and Transfer to PFF-101. The concept of unit procedures enables the user to model batch processes in great detail. A unit procedure is represented with a single equipment-looking icon on the screen. Multiple procedures can share the same equipment item as long as their cycle times do not overlap.

Operations: For every operation within a unit procedure, the simulator includes a mathematical model that performs material and energy balance calculations. Based on the material balances, it performs equipment-sizing calculations. If multiple operations within a unit procedure dictate different sizes for a certain piece of equipment, the software reconciles the different demands and selects an equipment size that is appropriate for all operations. In other words, the equipment is sized so that it is large enough that it will not be overfilled during any operation, but it is no larger than necessary (in order to minimize capital costs). In addition, the software checks to ensure that the vessel contents will not fall below a user-specified minimum volume (e.g., a minimum stir volume) for applicable operations.

The initialization of operations is done through appropriate windows. For instance, Fig. 2.25 shows the Oper.Cond's tab of a charge operation. Through this, the user specifies either the process time (duration) of the operation or the charge rate (based on mass or volumetric flowrate), and the program uses that information to calculate the duration. A third option is to set the duration of an operation equal to the duration of another operation or equal to the sum of durations of some other operations (through the "Set by Master-Slave Relationship" interface). The Emissions tab is used to specify parameters that affect emissions of volatile organic

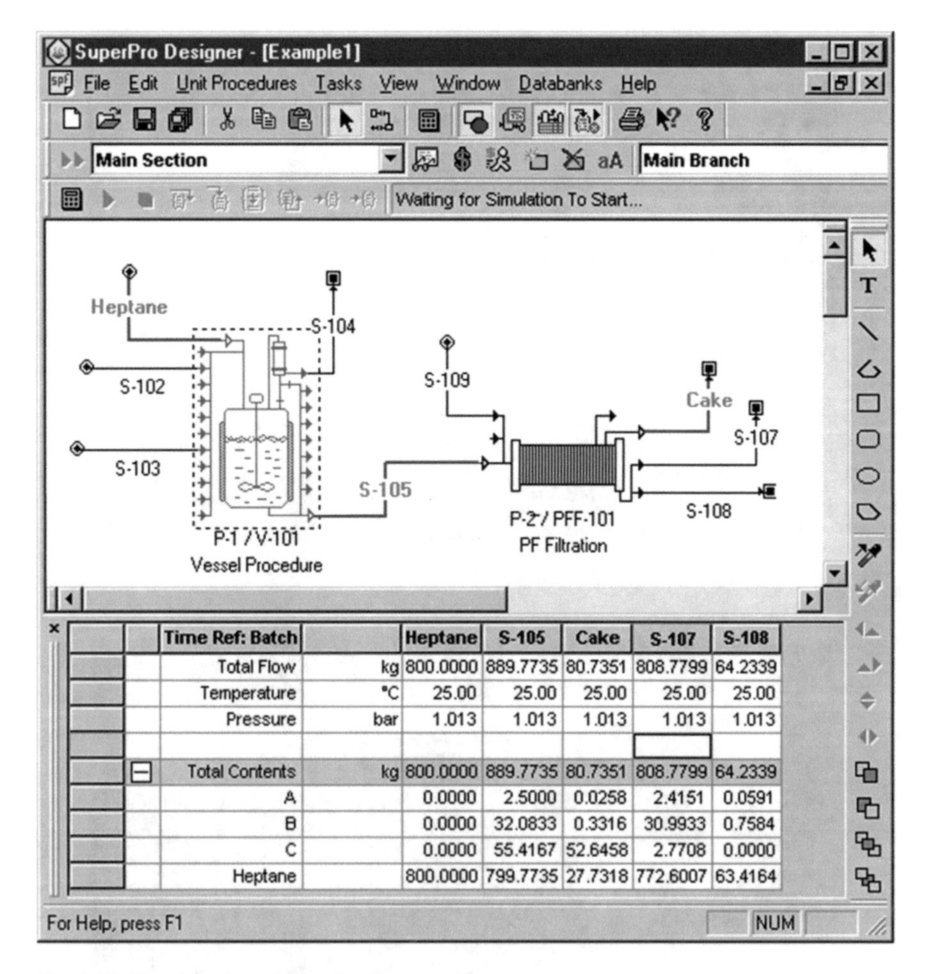

Fig. 2.23 User interface of SuperPro Designer®

compounds (VOCs). The Labor tab is used to specify the labor requirement for this operation. The Description tab displays a description of the process generated by the model (e.g., Charge 1000 L of Water at a rate of 150 L/min using stream Water-A). The user has the flexibility to edit the description and enter his/her own comments for documentation purposes. The Scheduling tab is used for specifying the Start Time of this operation relative to other events (e.g., the beginning of the batch, the beginning or end of some other operation in the same or a different procedure, etc.). SuperPro Designer® includes more than 120 operation models.

Component and Mixture Databases: The registration of pure components and mixtures is something that typically precedes the initialization of operations. SuperPro Designer® is equipped with two component databases, its own of 600 compounds and a version of DIPPR that includes 1700 compounds (the DIPPR

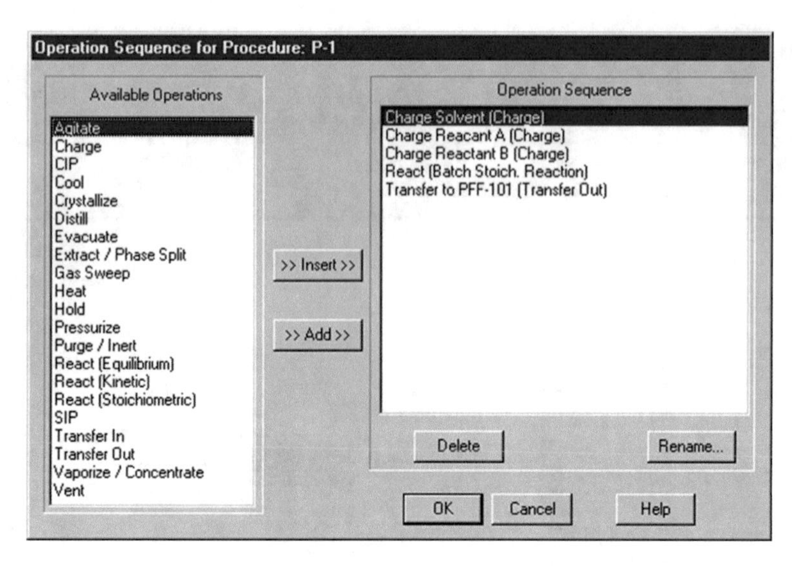

Fig. 2.24 Operation Sequence for Procedure

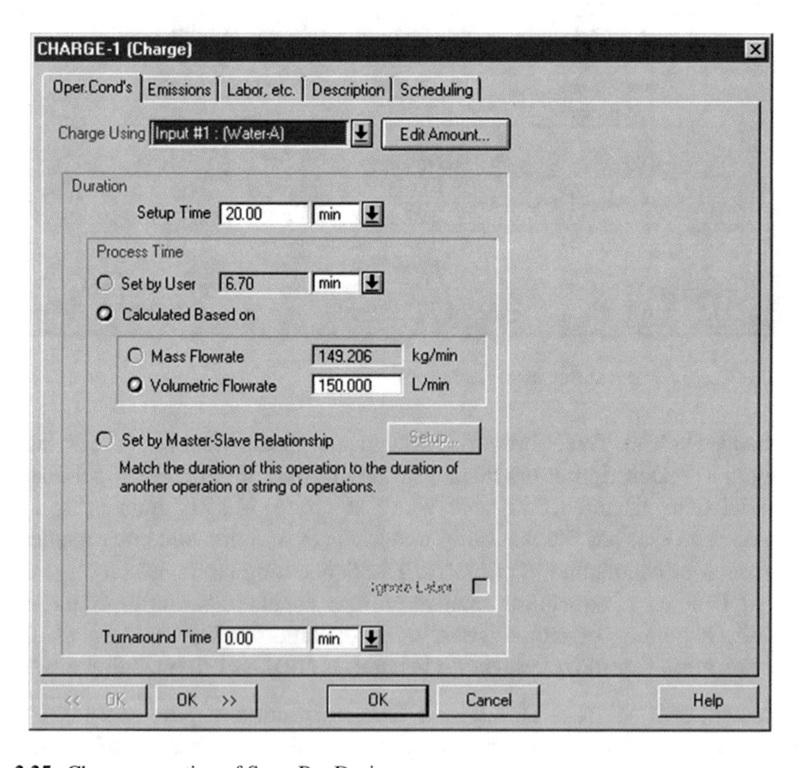

Fig. 2.25 Charge operation of SuperPro Designer

database must be purchased separately from Brigham Young University of Utah). It also comes with a user database where modified and newly created compounds can be saved. All database files are in MS Access format. Furthermore, SuperPro Designer® comes with mixture databases to represent buffers and other solutions that are commonly used in the biotech and other industries. Again, the user has the option to create his/her mixtures and save them in the user database.

For each pure component, the SuperPro Designer® databank includes thermodynamic (e.g., molecular weight, critical pressure and temperature, acentric factor, vapor pressure, density, specific heat, particle size, etc.), environmental (e.g., biodegradation data, octanol to water distribution ratio, Henry's law constant, component contribution to TOC, COD, BOD5, TSS, etc.), cost (e.g., purchasing price, selling price, etc.), and regulatory (e.g., type of pollutant) data.

2.5 PRO/II® Process Engineering

PRO/II® Process Engineering optimizes plant performance by improving process design and operational analysis and performing engineering studies (software.schneider-electric.com). This software was designed to perform rigorous heat and material balance calculations for a wide range of chemical processes. PRO/II® Process Engineering offers a wide variety of thermodynamic models to virtually every industry. PRO/II® Process Engineering is cost-effective, thereby decreasing both capital and operating costs.

PRO/II® is now available via the cloud in addition to the traditional on-premise access method. This cloud access has not only many benefits over on-premise access but also over other products with cloud access due to platform technology developed with simulation users in mind. PRO/II® has the following advantages:

- A secure user access control that allows the administrator to add and delete users or edit privileges as needed
- Simplify IT Overhead with the use of the product on pure on-demand cloud machines via a secure URL with no need for installation
- Seamless maintenance with new versions available as soon as they are released
- Flexible Usage and Pricing with SaaS business model based on minimum usage subscription and flexible, incremental usage credits
- Computer-based introductory training included

2.6 UniSim® Design Suite

Honeywell's UniSim® Design Suite is a process modeling software that provides steady-state and dynamic process simulation in an integrated environment (honeywellprocess.com). It provides powerful tools to help engineers evolve process optimization designs with lower project risks, prior to committing to capital expenditures. Some applications in process modeling using UniSim® Design Suite include:

- Process flowsheet development
- Utilizing case scenarios tool to optimize designs against business criteria
- Equipment rating across a broad range of operating conditions
- Evaluating the effect of feed changes, upsets and alternate operations on process safety, reliability, and profitability
- Accurately sizing and selecting the appropriate material for blowdown systems
- Monitoring equipment performance against operating objectives

2.7 gPROMS® ProcessBuilder

gPROMS® ProcessBuilder is an advanced process modeling environment for optimizing the design and operation of process plants (psenterprise.com). ProcessBuilder combines industry-leading steady-state and dynamic models with all the power of the gPROMS equation-oriented modeling, analysis, and optimization platform in an easy-to-use process flowsheeting environment (Fig. 2.26).

2.8 Process Simulation Exercises

With the purpose of offering the reader the opportunity to put into practice the knowledge acquired in this chapter, the following exercises are proposed:

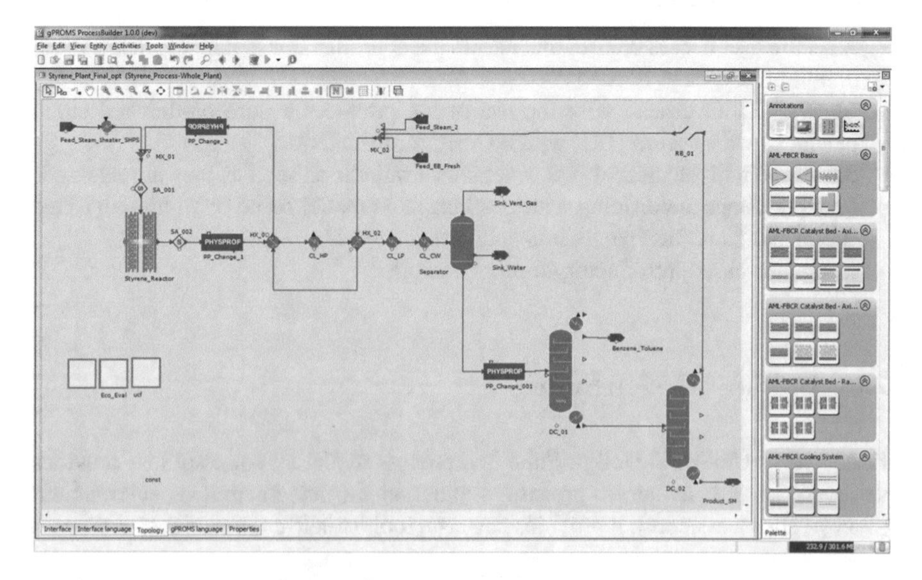

Fig. 2.26 User interface of gPROMS® ProcessBuilder

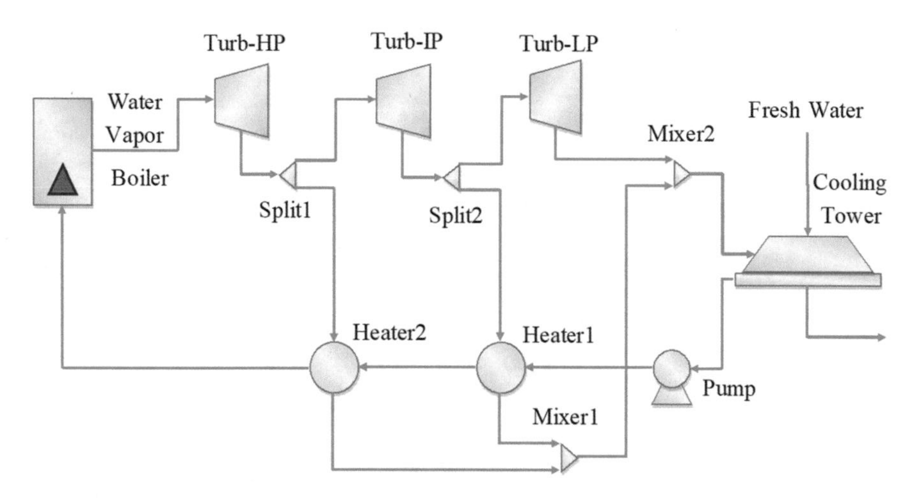

Fig. 2.27 Flowsheet of the regenerative Rankine cycle

1. Implement in Aspen Plus® the process flowsheet of a conventional Rankine cycle just as shown in Sect. 2.2 of this chapter (Fig. 2.12) with the following specifications:

   (a) Change the operating conditions in the boiler with a temperature equal to 500 °C and pressure of 50 atm (the discharge pressure of the pump must be of 50 atm too). What happened with the value of the ELECTR stream? Why?
   (b) Change the total flow rate of the WATER stream, with a value of 30,000 kg/h. What happened with the value of the ELECTR stream? Why?

2. Implement in Aspen Plus® the process flowsheet of a regenerative Rankine cycle, which is shown in Fig. 2.27, using the following operating conditions: a temperature of 580 °C, pressure of 38 atm, and a total flow of 1000 ton/day for the boiler output stream. The hot stream outlet temperature from the condenser is equal to 100 °C, hot stream temperature decrease of 10 °C in the first preheater and 100 °C in the second one, pressure of the pump of 40 atm, temperature of 600 °C, and pressure of 40 atm in the boiler. The split fraction is 0.8 in both splitters, and the pressure decreases are 20, 10, and 5 atm in the HP, LP, and LP turbines, respectively. Explain the results from this simulation.

2.9 Nomenclature

COM	Component Object Module
MS	Microsoft®
OLE	Object Linking and Embedding
VBA	Visual Basic for Application

Chapter 3
Metaheuristic Optimization Programs

There are optimization processes of industrial interest that involve functions that present a large number of local solutions, and therefore it is very difficult to determine the optimal solution using deterministic optimization techniques. For example, consider the case shown in Fig. 3.1, in which the cost function for the design of a heat exchanger relative to pressure drops is represented in a diagram. In this case, we can see that there are two local solutions, so if we use a local search procedure, the algorithm could be trapped in the solution that does not present the minimum cost since it complies with the stopping criteria of these optimization algorithms.

To solve these problems, we have proposed stochastic search algorithms based on natural phenomena such as simulated annealing (Kirkpatrick et al. 1983) and genetic algorithms (Goldberg 1989). These algorithms allow to search for problems that present a large number of local solutions as complex as the one shown in Fig. 3.2.

In this chapter, we present the general structure of simulated annealing (SA) and genetic algorithms (GA) as stochastic algorithms (other stochastic search algorithms can be seen in Fig. 3.3). These algorithms are applied directly to unrestricted problems where the problem consists of Min$f(\mathbf{x})$, with limits for the variables, $a \leq \mathbf{x} \leq b$.

3.1 Simulated Annealing

Simulated annealing (SA) is a metaheuristic technique based on an analogy of metal annealing. To describe this phenomenon, first consider a solid with a crystalline structure that is heated to melt, and then the molten metal is cooled to solidify again. If the temperature decreases rapidly, irregularities appear in the crystalline structure of the cooled solid, and the energy level of the solid is much

J. M. Ponce-Ortega, L. G. Hernández-Pérez, *Optimization of Process Flowsheets through Metaheuristic Techniques*,
https://doi.org/10.1007/978-3-319-91722-1_3

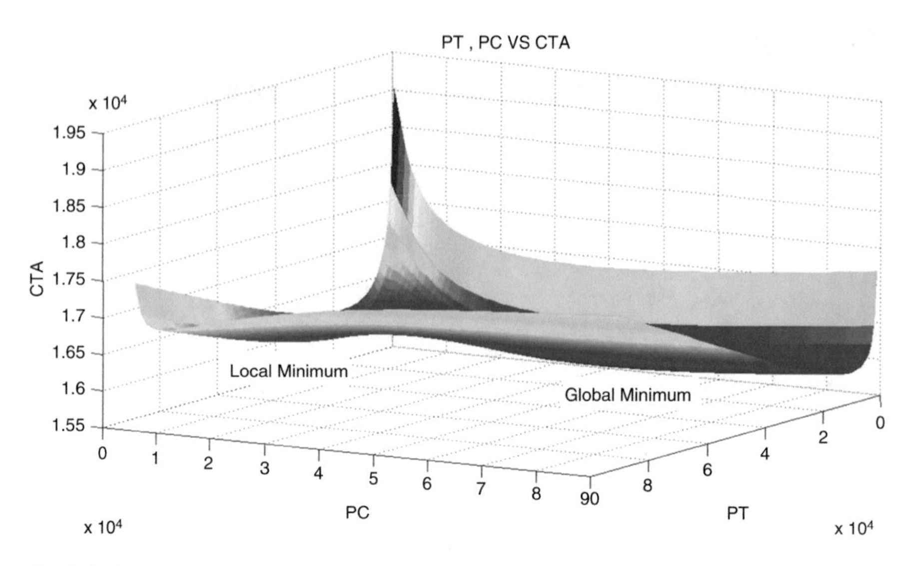

Fig. 3.1 Cost versus pressure drop graph for a particular case of a heat exchanger

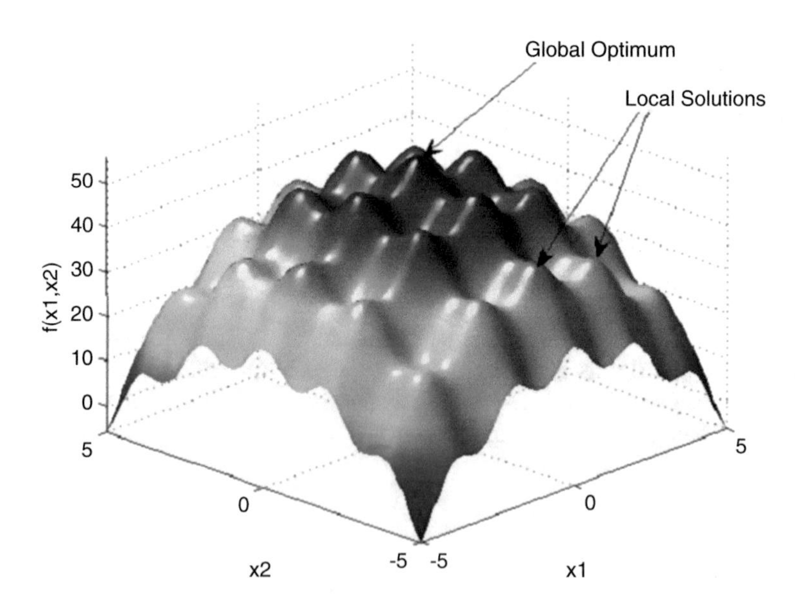

Fig. 3.2 Example of a function in which stochastic algorithms can be applied to find the global optimum

greater than a perfectly crystalline structure. If the material is cooled slowly, the energy level will be minimal. The state of the system at any temperature level is described by the coordinate vector **q**. At a given temperature, while the system remains in equilibrium, the state changes randomly, but the transition to states with lower energy level is more likely at low temperatures.

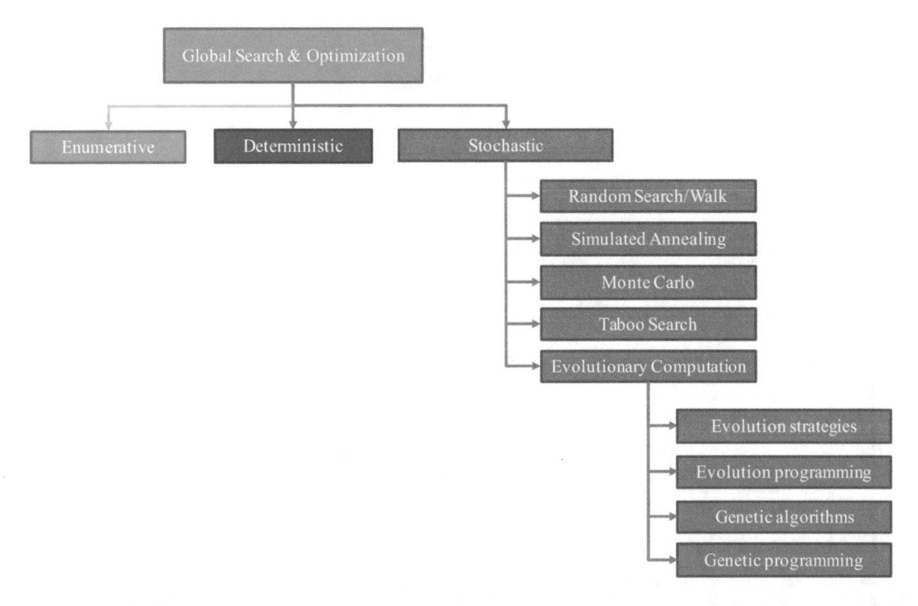

Fig. 3.3 Classification of stochastic search algorithms

To apply these ideas to a general optimization problem, we will designate the system state **q** as the optimization goal denoted as **x**. The energy level corresponds to the objective function, $f(\mathbf{x})$. The basic steps of the optimization algorithm SA are shown below (see Fig. 3.4):

0. Select a set of initial values for the search vector **x**, an initial temperature T, a lower limit for the temperature T^{low}, and a limit for the maximum number of iterations L.
1. Make $k = 0$.
2. Make $k = k + 1$.
3. Randomly select values for the unknown vector $\mathbf{x}'$.
4. If $f(\mathbf{x}') - f(\mathbf{x}) = 0$, make $\mathbf{x} = \mathbf{x}'$.
5. If $f(\mathbf{x}') - f(\mathbf{x}) > 0$, make $\mathbf{x} = \mathbf{x}'$ with a probability of exp-$(f(\mathbf{x}') - f(\mathbf{x}))/T$.
6. If $k < L$, return to step 2; otherwise continue with step 7.
7. Reduce the temperature $T = cT$, where $0 < c < 1$.
8. If $T > T^{low}$, returns to step 1; otherwise terminate the search procedure.

SA depends on the random strategy to diversify the search. The basic SA algorithm uses the Metropolis criterion to accept a motion. In this sense, the movements in which the objective function decrease are always accepted, whereas the movements that increase the objective function are accepted but with a probability of exp $(f(\mathbf{x}') - f(\mathbf{x})/T)$. When T approaches zero, the probability of accepting a movement in which the objective function increases is zero. Thus, when the temperature is high, many movements in which the objective function increases are accepted; in this way the method prevents it from being trapped prematurely in a local solution.

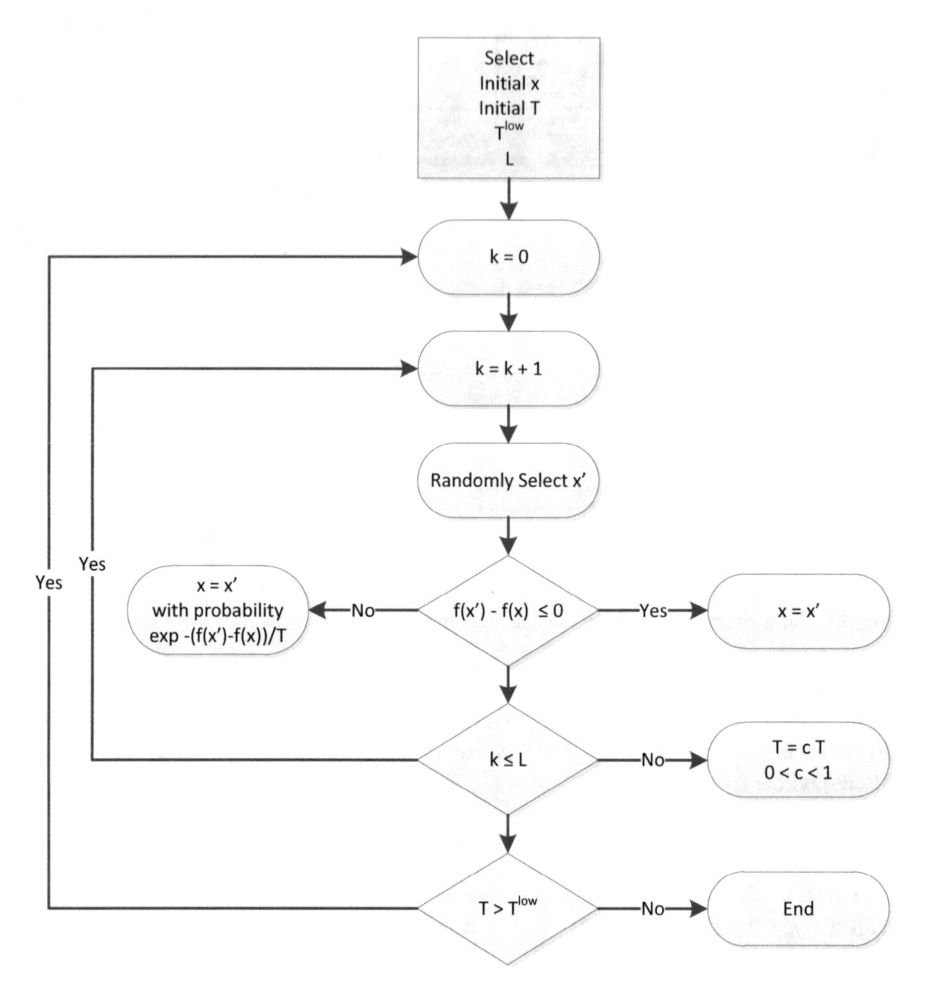

Fig. 3.4 General structure of the simulated annealing algorithm

3.2 Genetic Algorithms

Genetic algorithms (GA) are stochastic search techniques based on the mechanism of natural selection and genetics. GAs are particularly useful in non-convex problems or that include discontinuous functions. GAs differ from conventional optimization techniques because instead of having an initial solution (values for the optimization variable vector **x**), we have a set of solutions for the search vector called population. Each individual in the population is called a chromosome, which represents a solution to the problem. The chromosomes evolve through successive iterations which are called generations. During each generation, chromosomes are evaluated, using some form of measuring their abilities. To create the next generation, the new chromosomes are called descendants, which are created either by

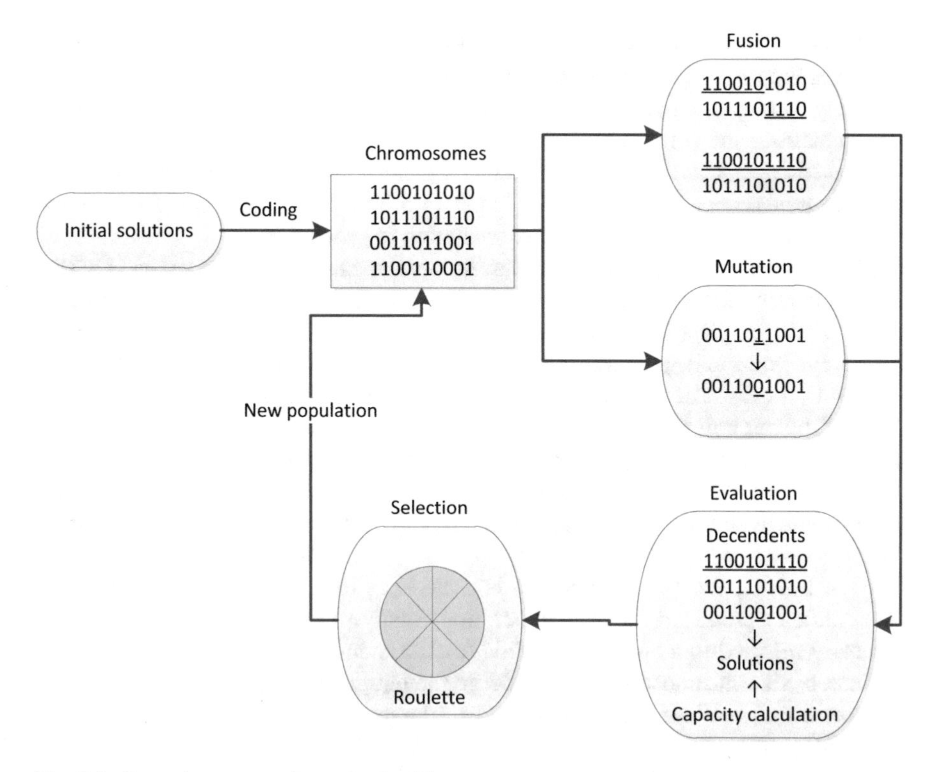

Fig. 3.5 General structure of genetic algorithms

combining two chromosomes of the current generation using the fusion operation or by modifying a chromosome at random using the mutation operation. The new generation is created by selecting, according to the value of their abilities, parents and descendants, and rejecting others to keep the size of the population constant. The more able chromosomes have a higher probability of being selected. After several generations, the algorithm converges to the best chromosome, which probably represents the optimal solution of the problem or a solution close to the optimal one. To explain this in detail, let us designate $P(t)$ and $C(t)$ as the parents and descendants of the current generation t. The general structure of the genetic algorithms (see Fig. 3.5) is described as follows:

0. Make $t = 0$.
1. Initialize $P(t)$.
2. Evaluate $P(t)$.
3. Combine $P(t)$ to produce $C(t)$.
4. Evaluate $C(t)$.
5. Select $P(t + 1)$ from $P(t)$ and $C(t)$.
6. Do $t \leftarrow t + 1$.
7. Finish if one of the convergence criteria is met; otherwise return to step 3.

Usually, the initial population is randomly selected. Recombination typically involves the fusion and mutation operations to produce the offspring. In fact, there are only two types of operations in genetic algorithms: (1) genetic operations (fusion and mutation) and (2) the evolution (selection) operation.

Genetic operators simulate the hereditary genetic process to create new offspring in each generation. The evolutionary operation imitates the process of Darwinian evolution to create populations from generation to generation.

Fusion is the main genetic operator. It operates on two chromosomes at a time and generates the offspring by combining the characteristics of two chromosomes. The simplest way to carry out the fusion operation is by randomly selecting a cut point and generating a descendant by combining the segment of one parent to the right of the cut point and the segment of the other parent of the left side of the cut point.

This method works correctly with the representation of the genes through bit strings, for example, with binary variables. The general behavior of genetic algorithms depends to a great extent on the effectiveness of the fusion operation used.

The fusion ratio (designated pc) is defined as the fraction of the number of descendants produced in each generation relative to the total population size (usually referred to as pop_size). This fraction controls the expected number of pc × pop_size chromosomes that undergo the merge operation. A high number of the fusion ratio allow the exploration of a larger solution space and reduce the possibility of installing in a false optimum. But if this ratio is too high, it could result in a waste of computation time in exploring non-promising regions of solution space.

Mutation is a fundamental operation, which produces spontaneous random changes in several chromosomes. A simple way to perform the mutation operation is to alter one or more genes. In genetic algorithms, the mutation operation plays a crucial role in (a) replenishing the lost genes of the population during the selection process so that they can be combined in a new context or (b) providing genes that were not considered in the initial population.

The mutation rate (designated by pm) is defined as the percentage of individuals in the population generated by mutation. The mutation rate controls the newly introduced genes in the population to be tested. If it is very low, many genes that could be useful will never be tested. But, if it is very high, there will be many random permutations, descendants will begin to lose their parent resemblance, and the algorithm will lose the ability to learn historically in the search process.

The convergence criteria for genetic algorithms are as follows: (a) if the maximum number of generations is exceeded, (b) if the maximum computation time is reached, (c) if the optimal solution is localized, or (d) if there are no improvements in successive generations or with respect to computation time.

3.2.1 Example of Codification

Consider the following problem:

$$\max f(x_1,x_2) = -x_1^2 - x_2^2 + 3\left[\cos(\pi x_1) + \cos(\pi x_2)\right]$$

subject to

$$-5 \le x_1 \le 5$$

$$-5 \le x_2 \le 5$$

Figure 3.2 shows a three-dimensional graph of the behavior of the objective function with respect to optimization variables, note the large number of local solutions associated with the problem.

To solve this problem, we first need to find a way to represent continuous optimization variables through binary strings. To perform this task, a strategy is to use a binary encoding of continuous variables. In this sense, we assume that the boundaries of the continuous variable x_j are defined in the interval $[a_j, b_j]$ and require a precision of decimal places; the mapping of that number can be represented by the following expression:

$$x_j = a_j + \text{decimal}\left(\text{subchain}_j\right) \times \frac{b_j - a_j}{2^{m_j} - 1}$$

where m_{j} represents the number of positions required by the subchain bit string to be able to represent the continuous variable x_j. To determine the value of m_j, we use the following expression:

$$2^{m_j - 1} < \left(b_j - a_j\right) \times 10^{d_j - 1} \le 2^{m_j} - 1$$

where d_j represents the number of positions after the decimal point needed to represent the continuous variable x_j. Thus, for our example, if we require three decimal places, we have the following for the variable x_1:

$$(5 - 0)10^2 = 500$$

then m_1 is equal to 9 (since $2^{9-1} < 500 \le 2^9 - 1$). For the variable x_2, m_2 is equal to 9 also since it is defined in the same interval. Therefore, the total number of bits needed to represent a chromosome in the example analyzed is equal to $m = m^1 + m_2 = 18$.

The decimal function (subchain_j) consists of the following summation:

$$\text{Decimal}\left(\text{subchain}_j\right) = \sum_m y_{m_j} 2^{m_j - 1}$$

Thus, for the continuous variable of the present example x_1, we have:

Position, m_1	9	8	7	6	5	4	3	2	1
Value, 2^{m1-1}	2^8	2^7	2^6	2^5	2^4	2^3	2^2	2^1	2^0
Binary, y^{m1}	0	1	0	1	0	1	0	1	0

In this case, the decimal (subchain$_1$) is equal to 170 and the variable x_1 is equal to −1.673. Note that if all binary variables are equal to one, we reproduce an upper limit (in this example 5). On the other hand, if all binary variables are equal to zero, we have the lower limit (in this example −5).

Thus, combining these two genes to represent x_1 and x_2, we form an 18-position chromosome representing a solution of the objective function:

$$v = \overbrace{010101010}^{x_1}\ \overbrace{101010101}^{x_2}$$

After explaining how to encode a binary variable through a binary string, we will randomly propose an initial population, considering a population size of five chromosomes as shown below:

$$\begin{aligned} v_1 &= [000110011 \quad 011001010] \\ v_2 &= [001101101 \quad 011010110] \\ v_3 &= [010110011 \quad 010001010] \\ v_4 &= [110110110 \quad 110111010] \\ v_5 &= [011101000 \quad 010010011] \end{aligned}$$

Whose corresponding decimal values are as follows:

$$\begin{aligned} v_1 &= [2.984, -1.751] \\ v_2 &= [2.123, -0.812] \\ v_3 &= [2.984, -1.751] \\ v_4 &= [-0.714, -1.340] \\ v_5 &= [-4.099, 2.866] \end{aligned}$$

Subsequently, we proceed with the evaluation of the capacities of each individual of the population through the objective function:

$$\begin{aligned} \text{eval}(v_1) &= f(2.984, -1.851) = 37.160 \\ \text{eval}(v_2) &= f(2.123, -0.812) = 45.117 \\ \text{eval}(v_3) &= f(2.984, -1.751) = 37.099 \\ \text{eval}(v_4) &= f(-0.714, -1.340) = 44.381 \\ \text{eval}(v_5) &= f(-4.099, 2.866) = 25.084 \end{aligned}$$

To produce the new generation, a number of elite individuals are selected to prevail as such in the next generation (this operation is known as elite count), in our example only one elite individual will prevail in the next generation, and this will be the one that presents a better adaptability (in our example this chromosome corresponds to v_2).

The other individuals from the next population will be generated through crossover and mutation operations. The percentage of individuals generated by fusion in our example will be 80%, and therefore the remaining 20% will be generated by mutation (in our case, we will have three descendants by fusion and one by mutation). To perform this task, we must first select the parents of the next generation. To perform this operation, the basic genetic algorithm uses the procedure known as roulette wheel, this operation is explained below. First, we calculate the total capacities of the current population:

$$F = \sum_{k=1}^{\text{pop\_size}} \text{eval}(v_k)$$

In our case, this value is $F = 188{,}843$. Then, we calculate the probability of each individual as follows:

$$p_k = \frac{\text{eval}(v_k)}{F}$$

In this case we have:

$$\begin{aligned} p_1 &= 0.196 \\ p_2 &= 0.238 \\ p_3 &= 0.196 \\ p_4 &= 0.234 \\ p_5 &= 0.132 \end{aligned}$$

Also, we need to determine the cumulative probability as follows:

$$q_k = \sum_{j=1}^{k} p_j$$

In this example, we have the following:

$$\begin{aligned} q_1 &= 0.196 \\ q_2 &= 0.435 \\ q_3 &= 0.632 \\ q_4 &= 0.867 \\ q_5 &= 1 \end{aligned}$$

Subsequently, two random numbers are generated in the interval [0,1] to select from the cumulative probability the parents of the first individual generated by fusion. Thus, if the first random number is 0.301, then the first parent will be v_2, since 0.301 falls between q_1 and q_2. In the same way, if the second random number is 0.453, the other parent will be v_3. Now, by applying the crossover operation, a cutoff point is generated in a random fashion, and we merge the two chromosomes to have:

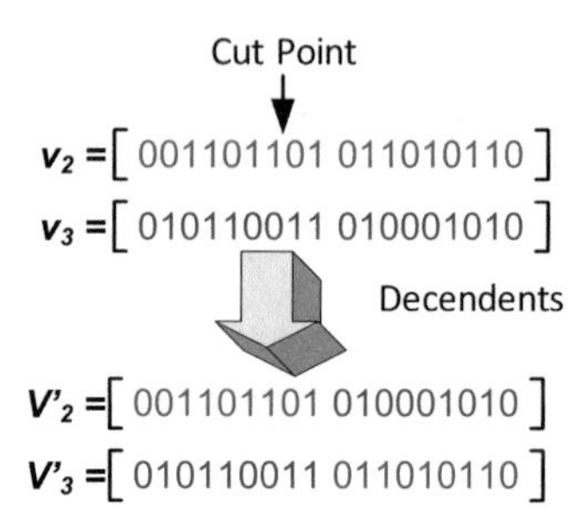

While the mutation operation involves making a random change in a gene on a chromosome. Thus, if in our example we generate a random number between 0 and 1 equal to 0.785, the selected chromosome to undergo a random change will be v_4, and the element to be changed is also randomly selected as shown below:

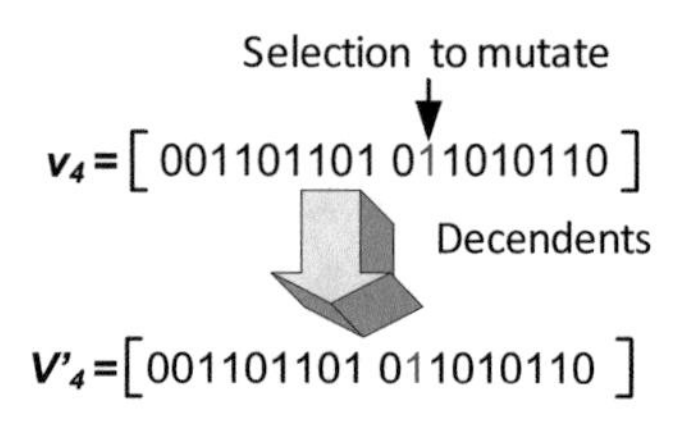

In this way, all descendants of the current generation are generated, and in turn these will be the parents of the next generation. The process is repeated, and after 102 generations we obtain the solution of the problem, in which the objective function is 56.00 with values for the optimization variables of $x_1 = 0$ and $x_2 = 0$ (point representing the overall optimal solution of the problem).

3.2.2 *Management of Restrictions*

The general procedure of genetic algorithms works correctly for problems of minimization or maximization of functions without restrictions. However, for handling additional constraints to the objective function, they must be treated in a special way. The most common strategy for the management of constraints through GA is to include a penalty function in the objective function, which includes the violation of the constraints within the objective function, and in this way the problem with constraints is transformed into an unrestricted problem by penalizing infeasible solutions. In general, there are two ways of representing the objective function including the penalty factor, one is by adding the objective function to the penalty term:

$$\text{eval}(\mathbf{x}) = f(\mathbf{x}) + p(\mathbf{x})$$

For a minimization problem, we have the following:

$$p(\mathbf{x}) = \begin{cases} 0 \text{ if } \mathbf{x} \text{ is feasible} \\ > 0 \text{ if } \mathbf{x} \text{ is infeasible} \end{cases}$$

The second strategy is to multiply the value of the objective function by the penalty term:

$$\text{eval}(\mathbf{x}) = f(\mathbf{x})\,p(\mathbf{x})$$

In this second case, we have for a minimization problem:

$$p(\mathbf{x}) = \begin{cases} 1 \text{ if } \mathbf{x} \text{ is feasible} \\ > 1 \text{ if } \mathbf{x} \text{ is infeasible} \end{cases}$$

There are several ways to determine the penalty term. For example, suppose that you want to solve the following problem:

$$\begin{aligned} &\min\ f(x) \\ &\text{suject to} \\ &g_i(x) \geq 0, \quad i = 1, 2, \ldots, m \end{aligned}$$

If we take the form of addition for the function to be optimized without restrictions, in this case the penalty term would be equal to:

$$p(\mathbf{x}) = \begin{cases} 0, \text{ if } \mathbf{x} \text{ is feasible} \\ \sum_{i=1}^{m} r_i g_i^2(\mathbf{x}), \text{ if } \mathbf{x} \text{ is feasible} \end{cases}$$

where r_i is a variable penalty coefficient for the restriction i.

Another way to handle a constrained optimization problem is shown below. Consider the following optimization problem:

$$\begin{aligned}&\min\ f(x)\\&\text{suject to}\\&g_i(x)\geq 0,\quad i=1,2,\ldots,m_1\\&h_i(x)=0,\quad i=m_1+1,\ldots,m\end{aligned}$$

The evaluation of the objective function as an unrestricted problem will be given by the following expression:

$$\text{eval}(\mathbf{x})=f(\mathbf{x})+p(t,\mathbf{x})$$

where

$$p(t,\mathbf{x})=\rho_t^{\alpha}\sum_{i=1}^{m}d_i^{\beta}(\mathbf{x})$$

here t is the iteration of GA. The penalty term is given as follows:

$$p(\mathbf{x})=\begin{cases}0, \text{ if } \mathbf{x} \text{ is feasible}\\|g_i(\mathbf{x})|, \text{ if } \mathbf{x} \text{ is infeasible to } 1\leq i\leq m_1\\|h_i(\mathbf{x})|, \text{ if } \mathbf{x} \text{ is infeasible to } m_1+1\leq i\leq m\end{cases}$$

$$\rho_t=Ct$$

where C is a constant.

There are several additional forms for penalty terms that can be studied in specialized sources; see, for example, Gen and Cheng (1997).

3.3 GA Toolbox of MATLAB®

First, make sure you have installed the toolbox for genetic algorithms. This toolbox contains the functions necessary to solve optimization problems through GA.

The first thing we need to do is an M file that must accept the vector of the independent optimization variables, and it returns a scalar representing the variable to be optimized.

To write an M file in MATLAB®, it is needed to follow these steps:

1. From the File menu, select New.
2. Select M File to open the file editor.
3. In the editor, write the function to be optimized, for example, presented in this chapter, we would have the following:

$$\text{function } z = f(x)$$
$$z = -(50 - x(1))2 + x(2)^\wedge 2 - 3(\cos(pi^* x(1)) + \cos(pi^* x(2)));$$

Here, z represents the objective function to be minimized, and x is the vector of optimization variables.

4. Record the f.m file in the working directory of MATLAB®.

There are two ways to use the GA in MATLAB®, one is through the command line and another is through the GA tool.

Firstly, we will explain how to solve the problem using the command line. The syntax of the MATLAB® GA routine from the command line is shown below:

$$[x \text{ fval}] = ga(@f,\text{nvariables},\text{options})$$

where

@f is the function that evaluates the capabilities of the chromosomes.
nvariables is the number of independent variables in the optimization function.
Options is an array that contains the options for the GA. If this argument is not included, the defaults are used.
x is the vector of optimization variables.
fval is the final value of the optimization function

To change the default settings in GA in MATLAB®, you must include the following instruction before calling the GA:

$$\text{Options} = \text{gaoptimset}(\text{“Parameter1,” value1, “Parameter2,” value2,}\ldots);$$

gaoptimset is a sub-routine that modifies the GA conditions used, and Parameter1 is the parameter to be modified and assigned the value value1. The same is true for Parameter2 and value2 and thus for all the parameters to be modified.

Some of the most important parameters that can be modified in the MATLAB® GA are shown in Table 3.1:

Additionally, MATLAB® has an application called GA tool. This application can be used to solve optimization problems with GA directly. To use the GA tool first, it is necessary to create an M file containing the objective function to be minimized. Afterward, open the GA tool working screen by entering the following command on the MATLAB® working screen gatool.

This will open a screen (see Fig. 3.6) where you first enter the function to be optimized (@f) in the fitness function field, as well as the number of optimization

Table 3.1 Options for GA in MATLAB®

Option	Parameter	Values	Description
Graphics options	"PlotFcns"	@gaplotbestf	Graph the best individual through successive generations
Population options	"PopulationType"	"doubleVector"	Continuous variables
		"bitstring"	Binary variables
	"PopulationSize"	number	Population size
	"InitialPopulation"	[ ]	Arrangement for the initial population
	"SelectionFcn"	@selectionstochunif	Uniform stochastic selection
		@selectionuniform	Uniform selection
		@selectionroulette	Roulette wheel selection
Reproduction	"EliteCount"	number	Number of best individuals remaining identical in the next generation
	"CrossoverFraction"	fractional number	Number of individuals reproducing by crossover
Stop criteria	"Generations"	number	Maximum number of generations
	"TimeLimit"	number	GA maximum computation time
	"FitnessLimit"	number	Limit value for the objective function
	"StallGenLimit"	number	Number of maximum permissible successive generations without improving the best chromosome
	"StallTimeLimit"	number	Maximum time allowed without improvements in the best chromosome
Display in screen	"Display"	""off"	Shows nothing
		""iter"	Displays the value of the target function each GA generation
		""final"	Shows the best individual in the final generation

variables. Subsequently, we can modify the default settings of the MATLAB® GA by modifying the parameters on the right side of the screen. Once you have all the conditions in which you want to run the GA for the specific problem, just press the start button, the algorithm is executed, and at the end it reports the results that we request.

In the same way, in the toolbox we can include minimum and maximum limits for the optimization variables in the bound lower and upper fields in the form of arrays (i.e., for two variables [number1 number2]). Linear constraints of equality and inequality can be introduced in matrix form across fields A and b for inequalities and Aeq and beq for equalities. Finally, nonlinear constraints are introduced through an M file.

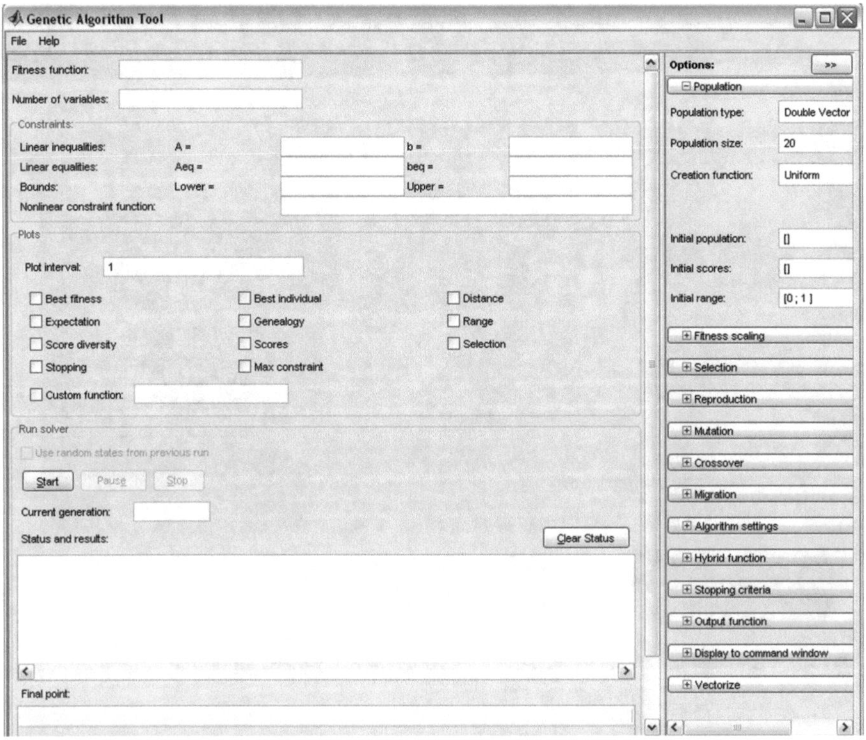

Fig. 3.6 GA toolbox of MATLAB®

3.4 EMOO Tool in MS Excel

Genetic algorithms (the flowchart is presented in Fig. 3.7) have proved to be a versatile and effective approach for solving optimization problems, but there are many situations in which the simple genetic algorithm does not perform particularly well, and various methods of hybridization have been proposed (Gen and Cheng 1997).

However, almost all these applications have been done using programs/platforms that are not readily used in the industry. On the other hand, engineers are familiar with MS Excel® and use it in both research and industrial practice (Sharma et al. 2012). Hence, an Excel-based multi-objective optimization (EMOO) program may represent a good alternative to develop stochastic optimization algorithms.

An improved multi-objective differential evolution algorithm with a termination criterion for optimizing chemical processes was developed by Sharma and Rangaiah (2013), which works with a termination criterion using the non-dominated solutions obtained as the search process. The multi-objective optimization hybrid method is, namely, the improved multi-objective differential evolution (I-MODE).

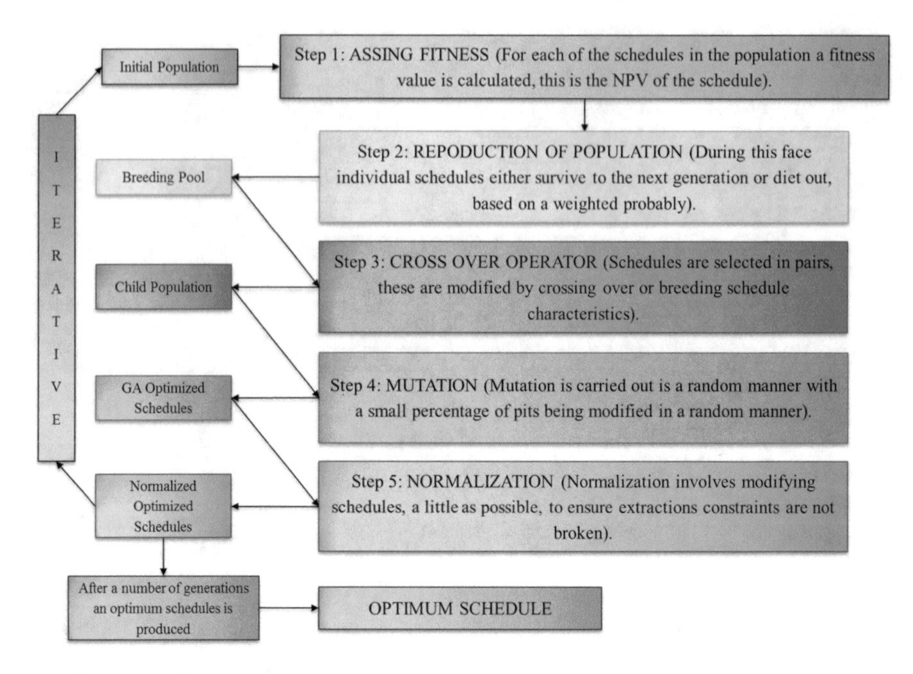

Fig. 3.7 Flowchart of simple genetic algorithm sequence of genetic algorithm

In the I-MODE (whose flowsheet is presented in Fig. 3.8), a population of NP individuals is randomly initialized inside the bounds of decision variables. Then, values of the objectives and constraints are calculated for each individual of the initial population. The taboo list size (TLS) is half of the population size, and taboo list (TL) is randomly filled with 50% individuals of the initial population; initial individuals are also identified as target individuals (i). A trial individual is generated for each target individual by mutation and crossover on three randomly selected individuals from initial/current/parent population. The elements of the mutant vector compete with those of the target vector, with a probability Cr to generate a trial vector. Taboo check is implemented in the generation step of the trial vector of I-MODE, and the trial individual is generated repeatedly until it is away from each individual in the TL by a specified distance called taboo radius (Tr). The Euclidean distance between trial individual and each individual in TL is calculated in the normalized space of decision variables for accepting the trial individual. After that, objectives and constraints are calculated for the temporarily accepted trial individual. The trial individual is stored in the child population and added to TL. After generating the trial individuals for all the target individuals of the current population, non-dominated sorting of the combined current and child populations followed by crowding distance calculation, if required, is performed to select the individuals for the next generation (G). The best NP individuals are used as the population in the subsequent generation.

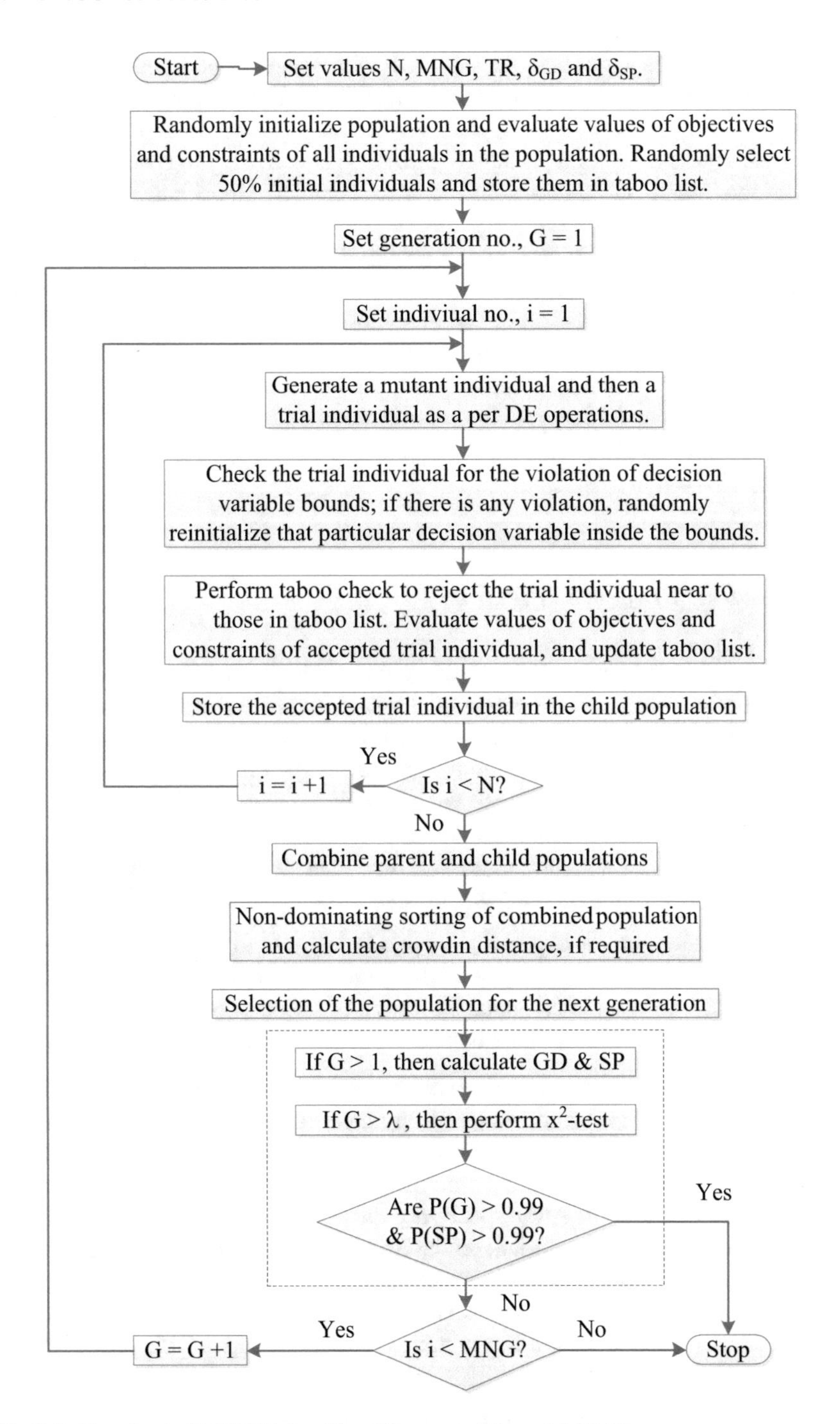

Fig. 3.8 Flowchart of I-MODE algorithm (Sharma and Rangaiah 2013)

For more details about how the I-MODE works and to obtain the optimization routine, you can consult the paper or contact the authors (Sharma and Rangaiah 2013).

For the optimization of a particular process, it is necessary to specify the values for the parameters associated with the used I-MODE algorithm, which are the following: population size (NP), number of generations (GenMax), taboo list size (TLS), taboo radius (TR), crossover fraction (Cr), and mutation fraction (F). In addition, it is necessary to select the decision variables and introduce the values for the lower and upper bounds expressed in homogenous units. All decision variables must be selected as a continuous or discontinuous variables, and the initial value for each is automatically the half between the minimum and the maximum possible values. The user interface of the optimization algorithm accepts inequality constraints, which can be introduced to run the optimization approach without any inequality constraint.

3.4.1 Main Program Interface

In order to give the reader the necessary knowledge to adequately manage the I-MODE, below are general notions of the user interface that use this stochastic algorithm in MS Excel®. For a detailed description of all the sheets that compose the I-MODE algorithm, it is recommended to review the user guide. Figure 3.9 shows the main program interface of the I-MODE.

As can be seen, the I-MODE algorithm has four fundamental parts in its main program interface, which are objective functions, design variables, inequality constraints, and algorithm parameters. Each of these sections will then be analyzed.

Objective Functions: As expected, this part is designed to introduce the objective function to be optimized by simply clicking the Add Objective Functions button. Next, a small window called Input Objective Function shown in Fig. 3.10 will appear where you have to fill three boxes.

The first box, which corresponds to Name, proposes a name for the function. In the second box, you are prompted to specify a cell value, that is, a cell where the target function is already specified. The last box corresponds to Goal, where you specify if you want to maximize or minimize the objective function, by default a minimization will appear. After specifying the information of the new objective function in each of the boxes, click on the Add button, and the specific information in the boxes will appear in the corresponding cell. If you want to cancel the introduction of a new click function on the Done button, this will erase all information previously entered in the boxes and will exit the Input Objective Function window.

Design Variables: In this part of the main program interface, you must enter the decision variables; it is the selected variables which will be manipulated in order

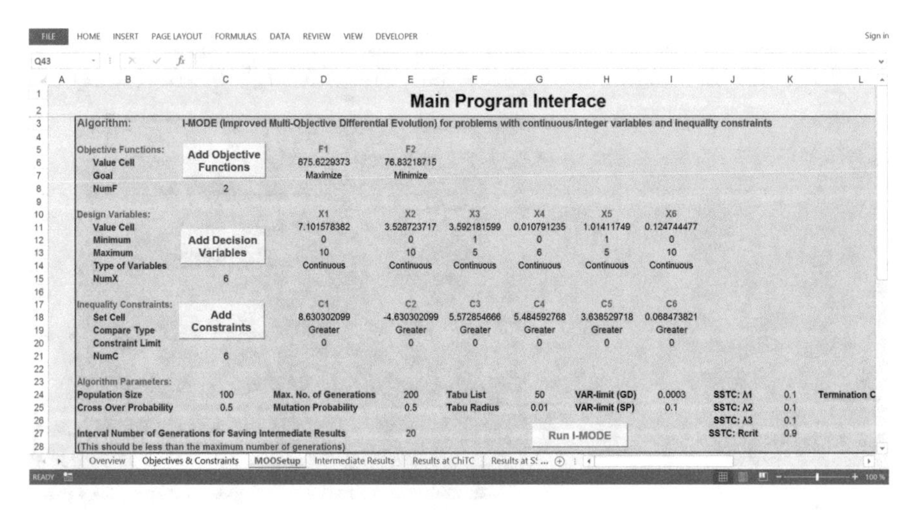

Fig. 3.9 Main program interface of I-MODE algorithm

Fig. 3.10 Input objective function of I-MODE algorithm

Input Objective Function

Input Objective Functions. (e.g. Maximize F1 taken from Cell "C12" from WorkSheet

Name (e.g. F1) F1

Value Cell 'Objectives'!C12

Goal Minimize

Add Done

to find the optimal point. To begin, you have to click the Add Decision Variables button, after which the window shown in Fig. 3.11 will appear.

As can be seen in Fig. 3.11, the input decision variables of I-MODE algorithm window contains four boxes where you must specify the name of the decision variable, the maximum and minimum values, as well as the type of variable, as appropriate. After entering the requested information, click on the Add button; otherwise to delete the information in the boxes, click on the Done button. The specific information in the boxes will appear in the corresponding cell.

Inequality Constraints: In the inequality restrictions, part of the main program interface of I-MODE algorithm, the constraints of the optimization problem must be

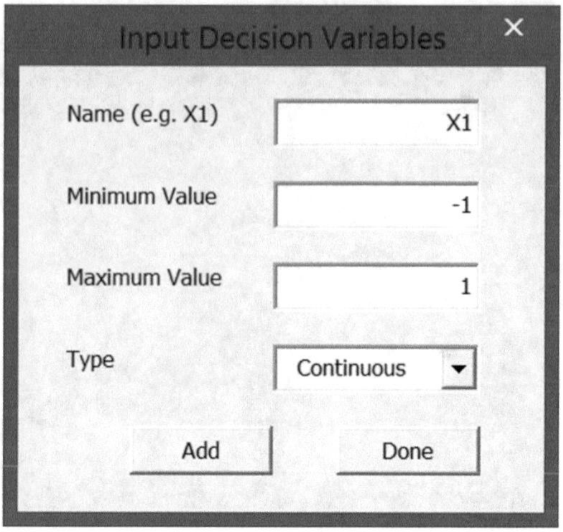

Fig. 3.11 Input decision variables of I-MODE algorithm

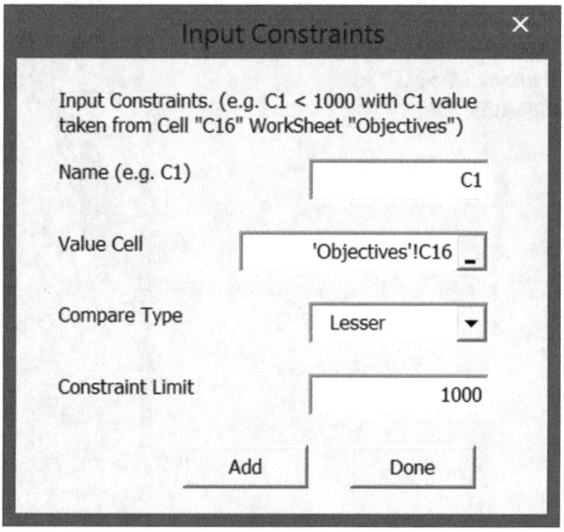

Fig. 3.12 Input constraints of I-MODE algorithm

specified, if any. To proceed, click the Add Constraints button after which the window shown in the Fig. 3.12 will be displayed.

The input constraints window contains four boxes in which you must enter the name that will be assigned to the constraint, the cell containing the constraint, the type of comparison, and the limit of the constraint. The information contained in the boxes structure the restriction, as shown by the example in the figure.

Algorithm Parameters: This is the part of the main program interface where it is necessary to specify the parameters associated with the used I-MODE algorithm, which are the following: population size (NP), number of generations (GenMax), taboo list size (TLS), taboo radius (TR), crossover fraction (Cr), and mutation fraction (F). All these values are entered directly into the corresponding

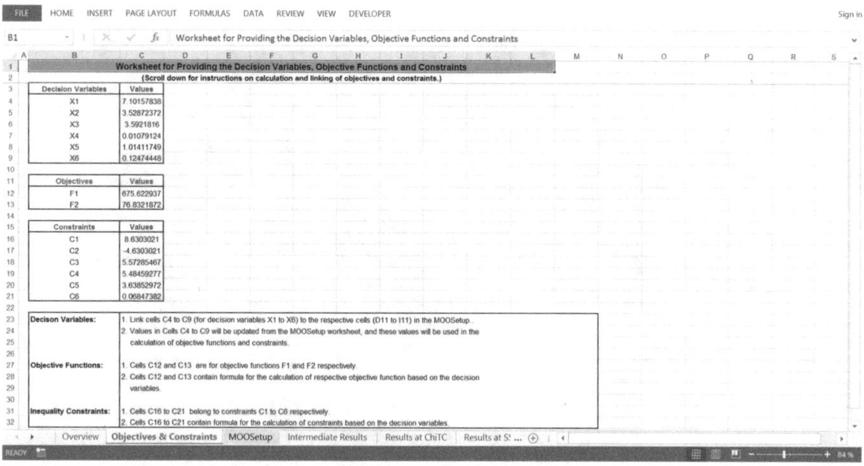

Fig. 3.13 Objectives and constraints of I-MODE algorithm

cell that indicates its name without needing to click on any button to add them. The numerical values of these parameters depend on the nature of the problem to be solved.

3.4.2 *Objectives and Constraints*

Many of the values that are entered in the main program interface of the I-MODE are referenced to another MS Excel sheet where they contain mainly the objectives and constraints of the problem to be optimized. In the objectives and constraints sheet shown in Fig. 3.13, you can identify the decision variables, objectives, and constraints. It is convenient to write in this part the numerical values of each aspect and to refer to them in the sheet of main program interface, this with the purpose of its easy modification in subsequent optimizations.

The value must be entered in its respective cell. It is important to emphasize that the number of decision variables, objective functions, and inequality constraints is not limited to the number proposed but can add more uncertainty of the cells of the usual way in MS Excel®.

3.5 Stochastic Optimization Exercises

1. Propose a chromosome to represent a string of bits for the variables x_1 and x_2. x_1 is defined in the interval [−10 10], while x_2 is defined in the interval [−20 20]. Accuracy of five decimal places is required.

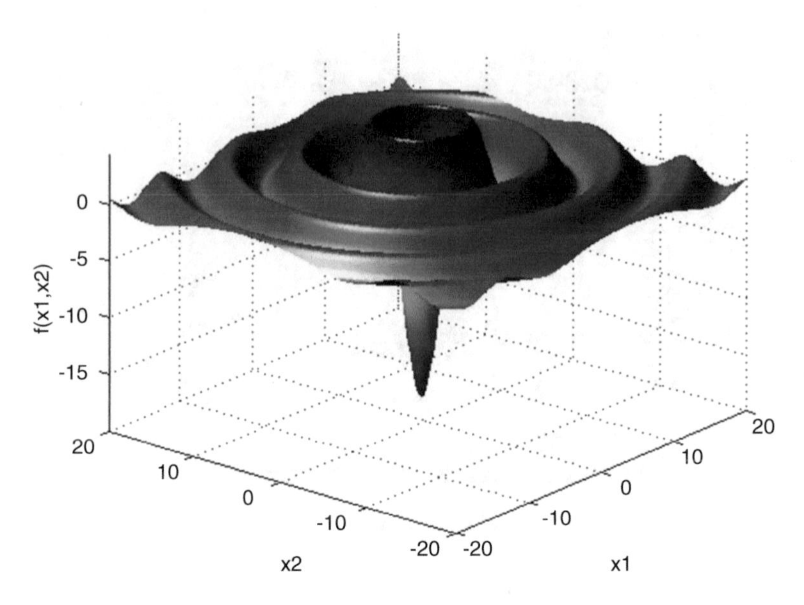

Fig. 3.14 Behavior of the function of problem 4

2. Propose a chromosome to model through a string of bits a set of variables given by x_1, x_2, and x_3 with an accuracy of six decimal places. The variables are defined in the following intervals:

   x_1 defined in [0–1]
   x_2 defined in [0.1 0.9]
   x_3 defined in [0–0.8]

3. For the example presented in the Section of GA, perform the calculations for the next ten generations, and see the behavior of the objective function as well as the evolution of the chromosomes. Plot the value of the objective function with respect to the generation number.
4. Consider the following problem:

$$\min f(x_1, x_2) = \frac{-20\sin\left(0.1 + \sqrt{(x_1 - 4)^2 + (x_2 - 4)^2}\right)}{0.1 + \sqrt{(x_1 - 4)^2 + (x_2 - 4)^2}}$$

   Figure 3.14 shows the behavior of the previous function:
   As can be seen in Fig. 3.14, this problem presents a large number of local solutions in which the deterministic algorithms could fail to locate the global optimal solution.
   Solve this problem using MATLAB® GA under the following conditions:

(a) For a population boundary of the optimization variables between [−20 20], with an elite count of 2 individuals, a fusion fraction of 0.8 and a limit for the infinite count time, as well as a maximum number of 50 generations.
First, solve the problem with a population size of five individuals. Then, solve the problem using a population size of 10 individuals, and finally solve the problem using a population size of 100 individuals.
Do you get the optimal solution in each of the population sizes? Explain the results.
(b) Now, change the maximum number of generations to 100. Then, solve the problem for 5, 10, and 20 individuals. In which cases do you get the optimal solution? Explain the results.
(c) For a population of 30 individuals and a maximum number of 100 generations, solve the problem with a value for the crossover fraction of 0.9, 0.5, and 0.2; what are the results? Explain the results obtained.
(d) Based on the results obtained in the previous sections, what would be the best values for GA parameters to solve this example?

5. Consider the following problem originally proposed by Hock and Schittkowski (1981):

$$\min f(x) = \sum_{j=1}^{10} x_j \left(c_j + \ln \frac{x_j}{x_1 + x_2 + x_3 + x_4 + x_5 + x_6 + x_7 + x_8 + x_9 + x_{10}} \right)$$

subject to

$$x_1 + 2x_2 + 2x_3 + x_6 + x_{10} = 2$$
$$x_4 + 2x_5 + x_6 + x_7 = 1$$
$$x_3 + x_7 + x_8 + 2x_9 + x_{10} = 1$$
$$x_i \geq 0.000001, \quad i = 1, \ldots, 10$$

where
$c_1 = -6.089, c_2 = -17.164, c_3 = -34.054, c_4 = -5.914, c_5 = -24.721, c_6 = -14.986,$ $c_7 = -24.100,\ c_8 = -10.708,\ c_9 = -26.662,\ c_{10} = -22.179.$
This problem includes a set of constraints that need to be included across a penalty term in the objective function. Solve the problem using the MATLAB® GA. The best solution to the problem is −47.760765; adjust the GA parameters to achieve this solution. What is your computation time?

6. Solve the following problem using the MATLAB® GA toolbox:

$$\min f(x) = \begin{cases} x\sin(x) - \cos(x), & 0 \leq x \leq 10 \\ x\ln(x), & 10 \leq x \leq 20 \\ x^2 \sin(x), & 20 \leq x \leq 30 \end{cases}$$

$$0 \leq x \leq 30$$

A special feature of this problem is that it includes discontinuous functions which are very difficult to manipulate with deterministic optimization methods.

7. Solve GA using the problem defined below:

$$\min\ f(x,y) = xy + y\sin(x) - x\cos(y)$$
$$-5 \le x \le 5$$
$$-5 \le y \le 5$$

   Determine the best parameters for GAs that allow finding the optimal solution in a shorter time.

8. The most commonly used type of heat exchanger is the shell-and-tube heat exchangers because they are robust units able to work in a wide range of pressure, flow, and temperature. Ponce-Ortega et al. (2009) developed an optimization approach based on genetic algorithms for the optimal design of this type of heat exchangers. This algorithm can be found in the following link:
   http://extras.springer.com
   Use this algorithm for solving the problems reported by Ponce-Ortega et al. (2009).
9. Process cogeneration is an effective strategy for exploiting the positive aspects of combined heat and power in the process industry. Bamufleh et al. (2013) developed an optimization framework for designing process cogeneration systems with economic, environmental, and social aspects. This algorithm can be found in the following link:
   http://extras.springer.com
   Use this code to solve the problems reported by Bamufleh et al. (2013).

3.6 Nomenclature

a_j	Lower bound for the interval of x_j
b_j	Upper bound for the interval of x_j
C	Population of decedents, set of solutions for the search vector
Cr	Probability
d_j	Number of positions after the decimal point needed to represent the continuous variable x_j
EMOO	Excel-based multi-objective optimization
f	Objective function
G	Generation
GA	Genetic algorithms
g_j	Set of inequality constraints where $i = 1, 2, \ldots, m_1$
h_j	Set of equality constraints where $i = m_1 + 1, \ldots, m$
i	Individuals

I-MODE	Improved multi-objective differential evolution
k	Number of iterations
L	Limit for the maximum number of iterations
m_j	Number of positions required by the subchain bit string to be able to represent the continuous variable
MODE-TL	Multi-objective differential evolution taboo list
MOO	Multi-objective optimization
NP	Population size
P	Population of parents, set of solutions for the search vector
pc	Fusion ratio
$\mathbf{q}$	Coordinate vector
r_i	Variable penalty coefficient for the constraint i
SA	Simulated annealing
t	Number of generation (iteration of GA)
T	Temperature
TL	Taboo list
T^{low}	Lower limit for the temperature
TLS	Taboo list size
Tr	Taboo radius
$\mathbf{x}$	Optimization variable vector
$\mathbf{x}'$	Unknown vector
x_j	Continuous variable
α	Penalty constant
β	Penalty constant
ρ	Penalty constant

Chapter 4
Interlinking Between Process Simulators and Optimization Programs

In this chapter, the methodology to achieve a successful link between the process simulator software and optimization programs using MS Excel® as a linker program will be described, as well as the direct linking of process simulator software with MS Excel® where it has been included a routine containing a stochastic optimization algorithm.

Recently, some approaches have been reported to optimize different processes based on process simulators and linked with different optimization approaches. For example, Segovia-Hernandez and Gomez-Castro (2017) reported an approach for optimizing chemical processes using stochastic optimization techniques and Aspen Plus®; Woinaroschy (2009) developed a simulation and optimization of citric acid production with SuperPro Designer using a client-server interface. Quiroz-Ramírez et al. (2017a) reported an optimization approach for selecting the feedstocks for biobutanol production considering economic and environmental aspects. Quiroz-Ramírez et al. (2017b) also reported a multi-objective stochastic optimization approach applied to a hybrid process production-separation in the production of biobutanol. Medina-Herrera et al. (2017) presented an optimal design of a multi-product reactive distillation system for producing silanes using a metaheuristic approach. Contreras-Zarazúa et al. (2016) presented a multi-objective optimization approach involving cost and control properties in reactive distillation processes to produce diphenyl carbonate. Santibañez-Aguilar et al. (2016) used a stochastic algorithm for designing biorefinery supply chains considering economic and environmental objectives. González-Bravo et al. (2016) developed a multi-objective optimization for dual-purpose power plants and water distribution networks.

It is noteworthy that there is not reported a general framework to link any process simulator to any metaheuristic optimization technique. This way, in this chapter is presented a general framework to link any process simulator to metaheuristic optimization approaches to optimize process flowsheets. The proposed approach implements a link between the process simulator (Aspen Plus®, Aspen HYSYS®, SuperPro Designer®, etc.) and an optimization algorithm (it can be a multi-objective

J. M. Ponce-Ortega, L. G. Hernández-Pérez, *Optimization of Process Flowsheets through Metaheuristic Techniques*,
https://doi.org/10.1007/978-3-319-91722-1_4

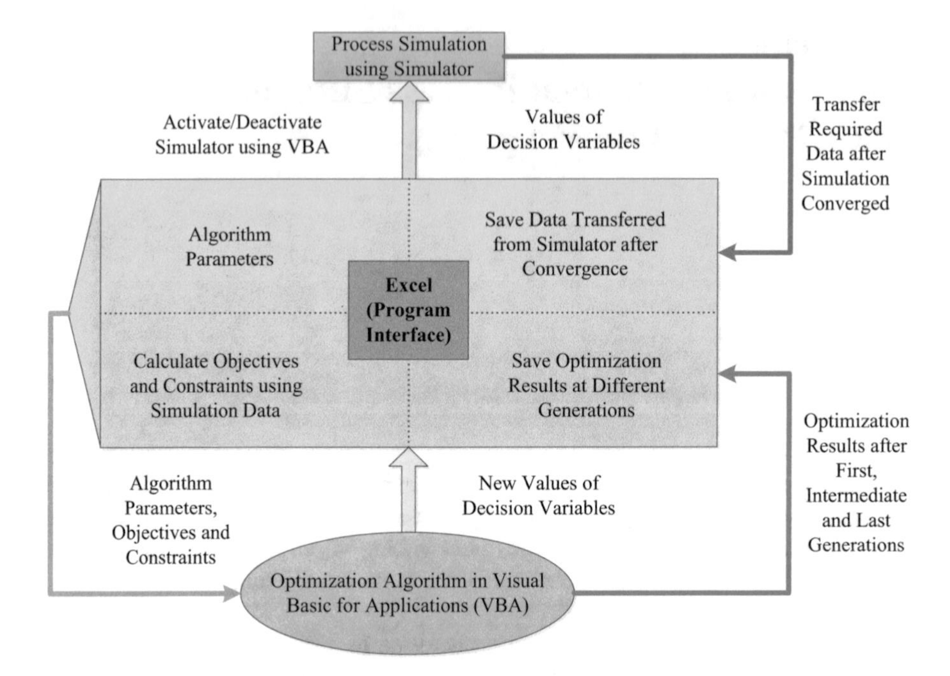

Fig. 4.1 Interface between a process simulator and Excel® (containing an optimization program)

differential evolution with taboo list implanted in MS Excel or other metaheuristic techniques implemented in other software), which can be linked through a COM software via Visual Basic for Applications (VBA).

The implementation of the global optimization approach involves a hybrid platform, which links simulator software and Microsoft Excel® through the implementation of a Component Object Module (COM) technology (the flowchart is presented in Fig. 4.1).

A client-server interface based on COM technology can be implemented. Using COM technology, it is possible to add code so that the applications behave as an Object Linking and Embedding (OLE) automation server. The use of the methods of this library to interoperate with other Windows applications (such as Excel®) requires the use of a common scripting language, and Visual Basic® for Applications (VBA) can be used in this case. An interface between Excel® and Aspen Plus®, based on COM technology, using Excel-VBA scripts (Birnbaum 2005) can be implemented.

During the optimization process, a decision vector of design variables is sent from Excel® to Aspen Plus®; for example, in this process simulator rigorous calculations for the data that identify a particular design of process are obtained (e.g., temperature and pressure in the boiler, split fraction in the splitter, etc.) via resolution of phase equilibrium along with the complete set of mass and energy balances. These data are returned from Aspen Plus® to Excel® for the calculation of objective functions; the values obtained for the objective functions are evaluated, and new vectors of design variables are generated according to the stochastic procedure of this method.

4.1 Previous Knowledge

4.1.1 MS Excel® Configuration

A very important aspect to take into account is that to achieve this type of linking requires access to the routines and optimization algorithms; for this it is necessary to configure the tabs of MS Excel® so that the "Developer" tab appears; since in the startup configuration of this MS Office® program this tab does not appear, it is necessary to customize the ribbon when it is first used, after which it is no longer necessary to do so again. For more details about how to configure the ribbon, it is recommended to the readers to see the appendices in this book.

Another important aspect to consider for the previous configuration in the MS Excel® program is in the section of MS VBA (Microsoft Visual Basic for Applications). To access to this part of the program, you have to click the button shown in Fig. 4.2.

Then, the window shown in Fig. 4.3 appears, and you can select the project that contains the optimization algorithm or the linking subroutine.

To adequately run the algorithm that contains the Aspen Plus® variable call, it is necessary to activate the libraries for this program. It is called Aspen library. For activating this, we must go to the Tools tab and the References option (see Fig. 4.4).

Later, we proceed to look for the Aspen Plus® libraries included, and we select all of them (see Fig. 4.5).

4.1.2 Object Name of the Simulator File

To do this, it is necessary to open the windows explorer in the folder containing the files needed for linking. Then, to check the exact route of the simulation, select the simulator software backup file and right click on it, and then a menu will be displayed where we will select the last option "Properties" (Figs. 4.6 and 4.7).

A window will open where we select the "Security" tab. Here, there are shown some data security of the file; you must copy the whole path of the box "Object Name" (Figs. 4.8 and 4.9).

After that, we open the routine that contains the stochastic optimization algorithm in Visual Basic that is inside the "Developer" tab. It is necessary to select the appropriate module that contains the optimization algorithm and that is located in the part of the linking subroutine. In this part, we declare the path of the simulation backup file by pasting the object name that we copied (Fig. 4.10).

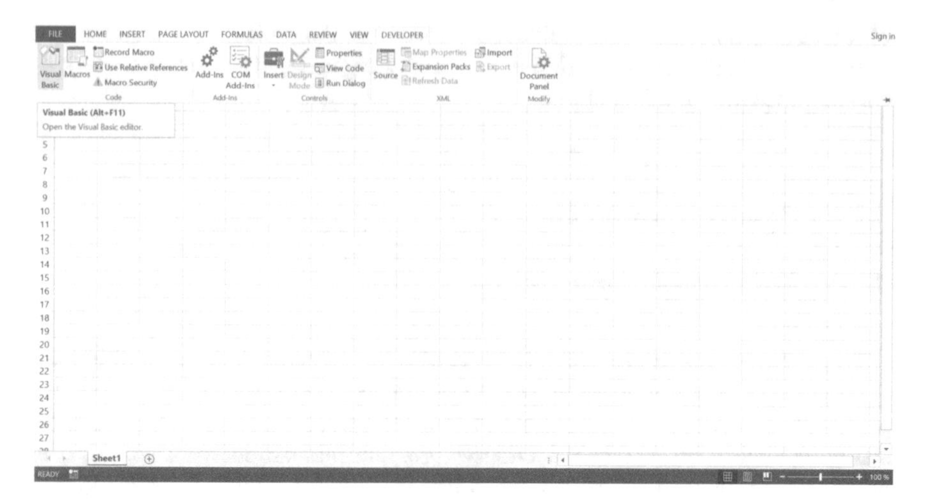

Fig. 4.2 Screenshot of the position of Visual Basic® button

Fig. 4.3 Screenshot of the MS Visual Basic for Applications

4.2 Link Between Aspen Plus® and MS Excel®

An interface between the process simulation software Aspen Plus® and MS Excel® can be established (as can be seen in Fig. 4.11). This type of direct communication between the process simulation software and the program that contains the optimization algorithm presents multiple advantages that can be reflected directly in computer resources such as the computation time.

For more information about how to implement this link, we recommend reviewing the tutorial video that is included as additional material for this book. You can access it using the following link:

http://extras.springer.com

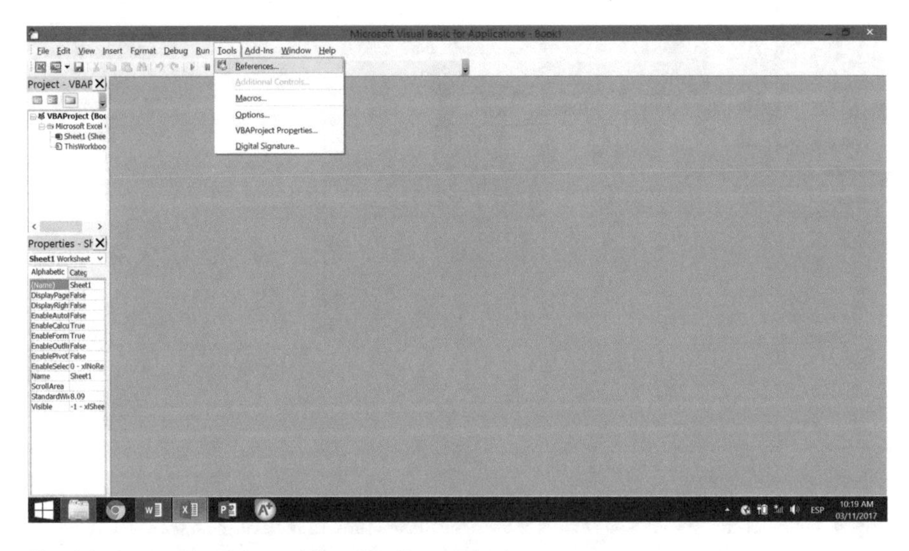

Fig. 4.4 Screenshot of the position of References button

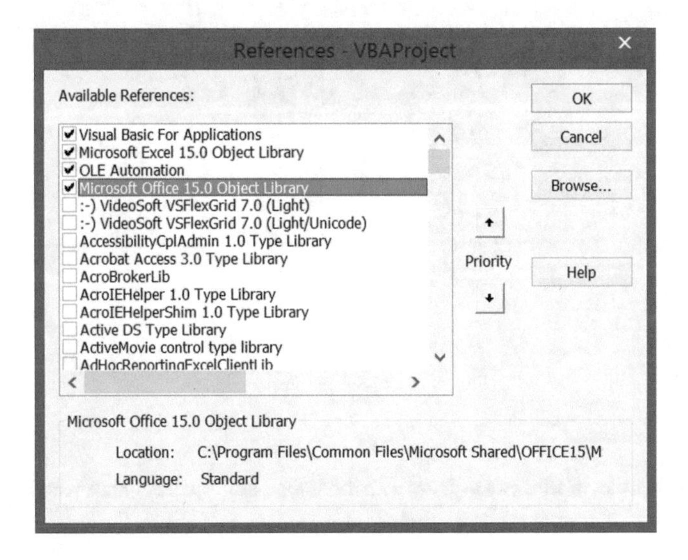

Fig. 4.5 Screenshot of the References window

4.2.1 *Subroutine to Link Aspen Plus® and MS Excel®*

To perform the successful link between the process simulator software Aspen Plus® and a program that has the stochastic optimization tools using MS Excel® as a linker program, it is necessary to follow the steps described by the general syntax for the code used, which is shown in Fig. 4.12.

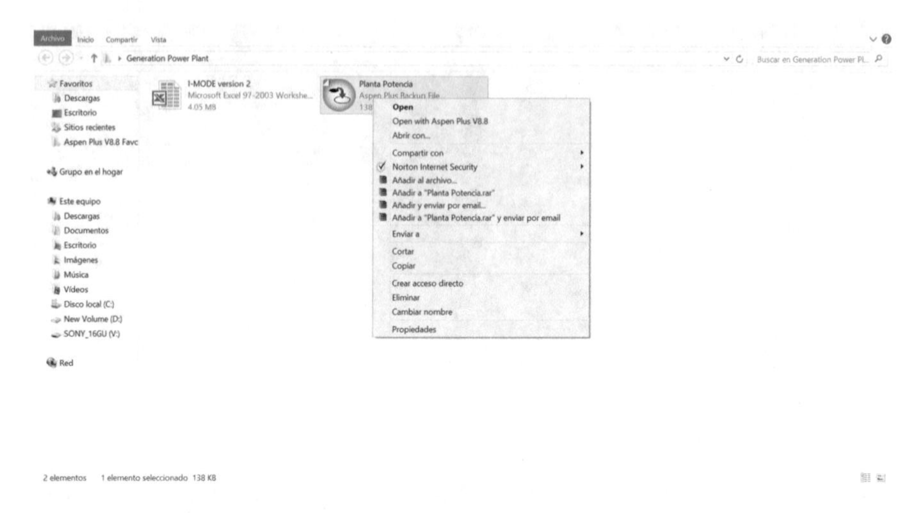

Fig. 4.6 Screenshot of windows explorer with the "Properties" option of the Aspen Plus® file

Fig. 4.7 Screenshot of windows explorer with the "Properties" option of the SuperPro Designer® file

4.2.2 Files to Link Aspen Plus® and MS Excel®

Before you begin, you need two files to perform the linking (shown in Fig. 4.13). The first one will be the backup file of Aspen Plus® that corresponds to the data of the simulation of processes previously elaborated. The second file will be from the MS Excel® linker program, which will include a routine that will call the variables of the simulator, placing input values and reading the response variables. For this direct linking, no other optimizer program file is required, since the stochastic optimization algorithm is within the MS Excel® linker program.

Figure 4.14 shows a screenshot of windows explorer with the two files needed to start linking: Aspen Plus® and MS Excel®. The orders in which these files appear

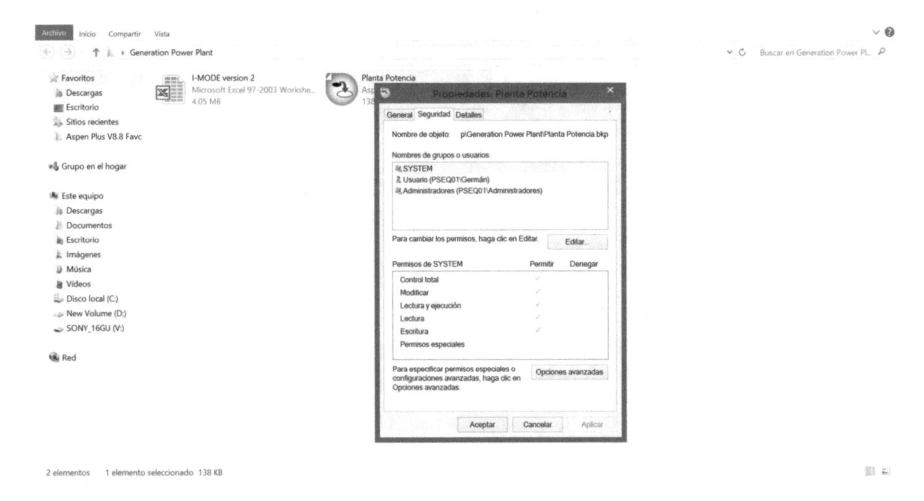

Fig. 4.8 Screenshot of windows explorer with the "Security" tab of the Aspen Plus® file where the "Object Name" can be seen

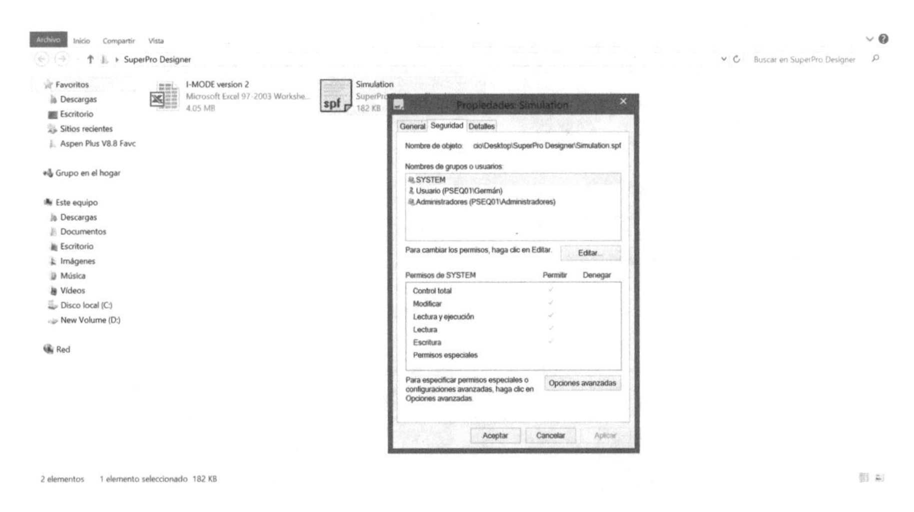

Fig. 4.9 Screenshot of windows explorer with the "Security" tab of the SuperPro Designer® file where the "Object Name" can be seen

from left to right are linker program with the optimization algorithm and backup simulator file.

Once it has been corroborated that you have the essential files for implementing the link, we can begin with the explanation of the steps to follow, which are described in the next section.

If the user does not have any files for the process simulation in Aspen Plus® and for the optimization algorithm in MS Excel®, the files shown in Fig. 4.14 can be found in the following link:

http://extras.springer.com

Fig. 4.10 Screenshot of linking subroutine where the simulation file route must be pasted

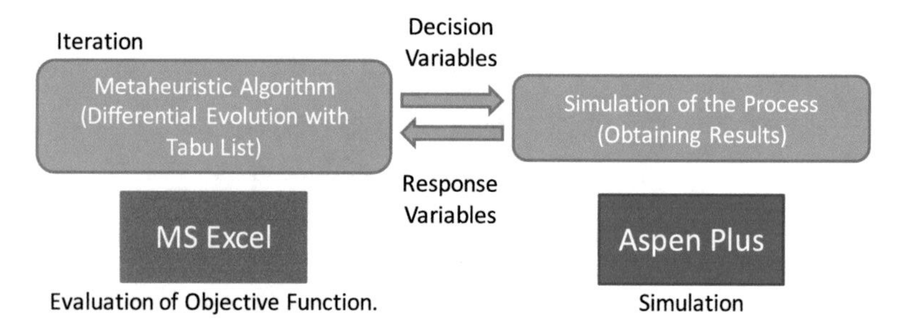

Fig. 4.11 Interface between Aspen Plus® and MS Excel®

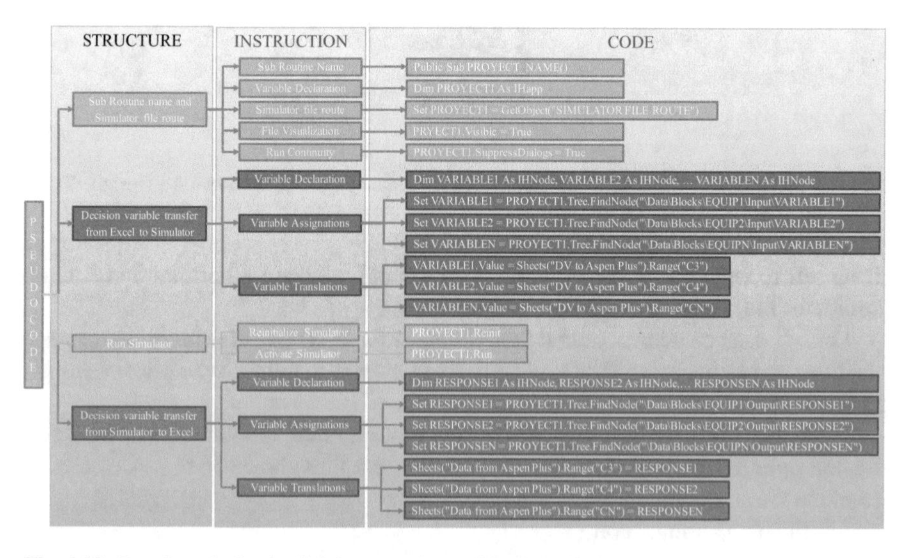

Fig. 4.12 Pseudo code for the link between Aspen Plus® and MS Excel®

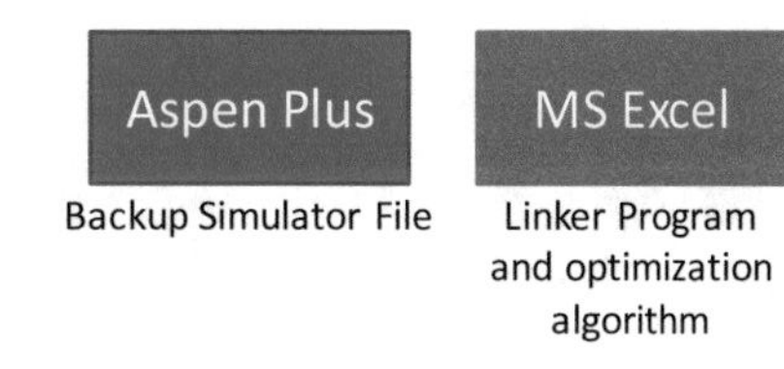

Fig. 4.13 Two files needed to start linking Aspen Plus® and MS Excel®

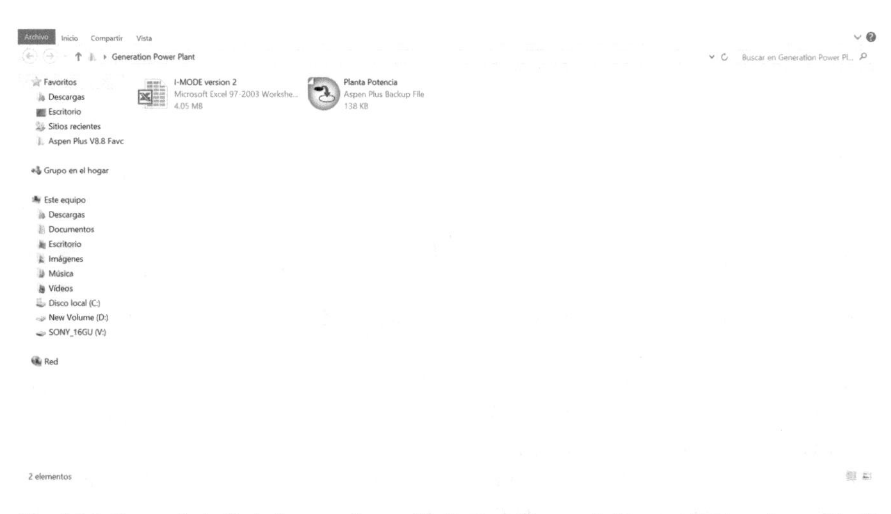

Fig. 4.14 Screenshot of windows explorer with the two files needed to start linking: Aspen Plus® and MS Excel®

4.2.3 *Call Name of Aspen Plus® Variables*

Once the path of the backup file of the process simulation has been declared, we proceed with the declaration of decision variables and response variables in the linker program. To do this, we need to open the file containing the simulation in Aspen Plus® and select the "Customize" tab and choose the "Variable Explorer" option (Fig. 4.15).

Then, you will see a tree of options where we proceed to find the name of the variable that uses the Aspen Plus® variable explorer and that corresponds to the process variable that we want to link. The user must look at the tree of the variable explorer, the variable that wants to be either as a decision variable or as a response variable. In order to find the desired variable, a logical sequence of search is followed according to the equipment or stream to which this variable belongs. For example, if the user is looking for the variable "Feeding temperature" of an equipment called "Boiler," the path to be followed in the variable explorer tree is as follows: Root > Data > Blocks > BOILER > Input > Temperat ure. Once the path has been followed to the desired variable, a series of data appears on the right-hand side of the variable explorer. There are many variables with a similar name, so one way to verify that it is the appropriate variable is to verify in the attribute "Value" effectively that variable has the numerical value

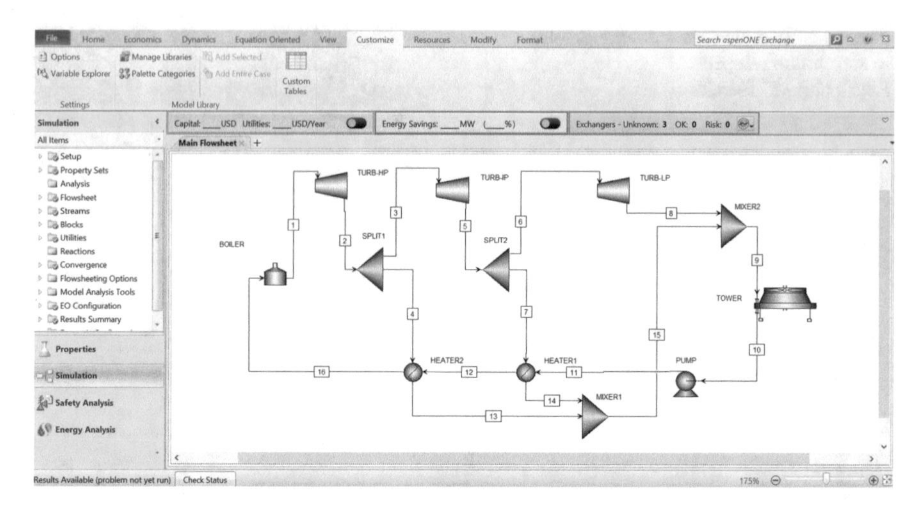

Fig. 4.15 Screenshot of variable explorer position in Aspen Plus®

that corresponds to it. Since we are sure that it is the searched variable, we copy the contents of the value corresponding to the "Call" attribute (Fig. 4.16).

Once you have copied the name of the variable with which the COM technology will make the call, proceed to open the link subroutine again, and look for the part in which the variables are assigned; pasted here is the variable name for the call (Fig. 4.17).

4.3 Link Between SuperPro Designer® and MS Excel®

An interface between the process simulation software SuperPro Designer® and MS Excel® can be established (as can be seen in Fig. 4.18).

For more information about how to implement this link, we recommend reviewing the tutorial video that is included as additional material for this book. You can access it using the following link:

http://extras.springer.com

4.3.1 Subroutine to Link SuperPro Designer® and MS Excel®

To successfully perform the link between the process simulator software SuperPro Designer® and a program that has the stochastic optimization tools using MS Excel® as a linker program, it is necessary to follow the steps described by the general syntax for the used code, which is shown in Fig. 4.19.

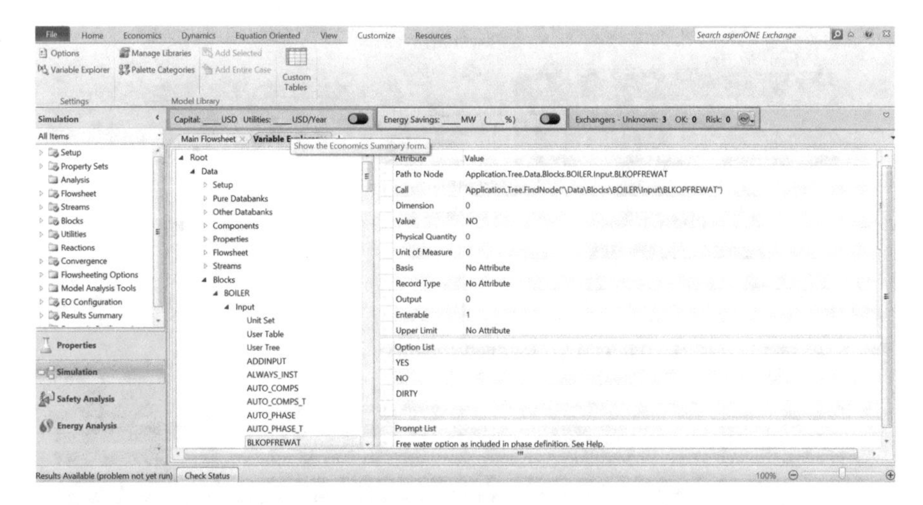

Fig. 4.16 Screenshot of variable explorer tree in Aspen Plus®

Fig. 4.17 Screenshot of linking subroutine where the variable name call must be pasted

4.3.2 Files to Link SuperPro Designer® and MS Excel®

Before you begin, you need two files to perform the linking (shown in Fig. 4.20). In the first one, there is implemented the backup file of SuperPro Designer® that corresponds to the data of the simulation of processes previously elaborated. The second file will be the MS Excel® linker program, which will include a routine that will call the variables of the simulator, placing input values and reading the response variables. For this direct linking, no other optimizer program file is required, since the stochastic optimization algorithm is within the MS Excel® linker program.

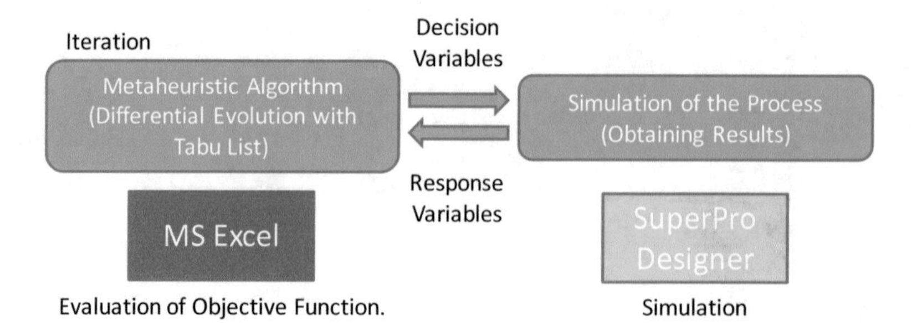

Fig. 4.18 Interface between SuperPro Designer® and MS Excel®

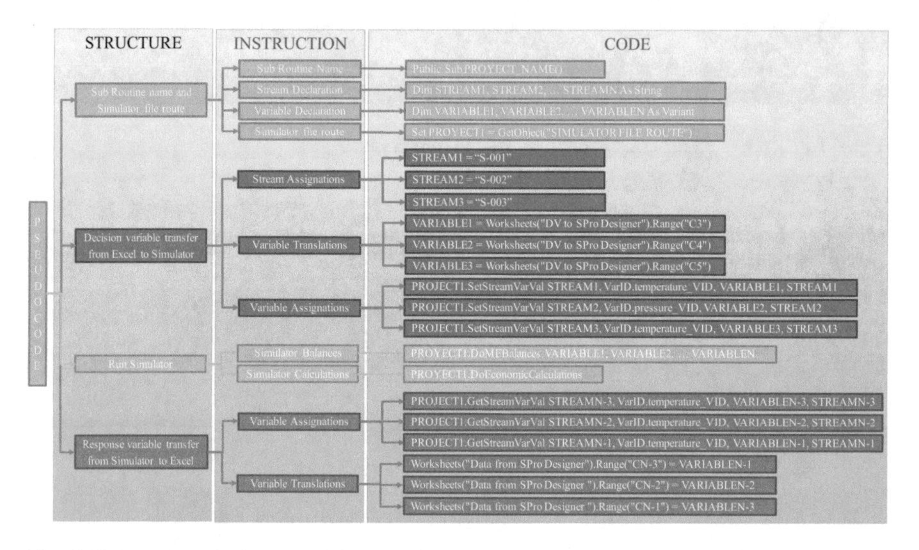

Fig. 4.19 Pseudo code for the link between SuperPro Designer® and MS Excel®

Fig. 4.20 Two files needed to start linking SuperPro Designer® and MS Excel®

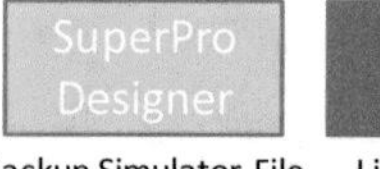

Figure 4.21 shows a screenshot of windows explorer with the two files needed to start linking SuperPro Designer® and MS Excel®. The orders in which these files appear from left to right are linker program with the optimization algorithm and backup simulator file.

Once it has been corroborated that you have the essential files for the link, we can begin with the explanation of the steps to follow, which are described in the next section.

If the user does not have any files for the process simulation in SuperPro Designer® and for the optimization algorithm in MS Excel®, the files shown in Fig. 4.21 can be found in the follow link:

http://extras.springer.com

Fig. 4.21 Screenshot of windows explorer with the two files needed to start linking SuperPro Designer® and MS Excel®

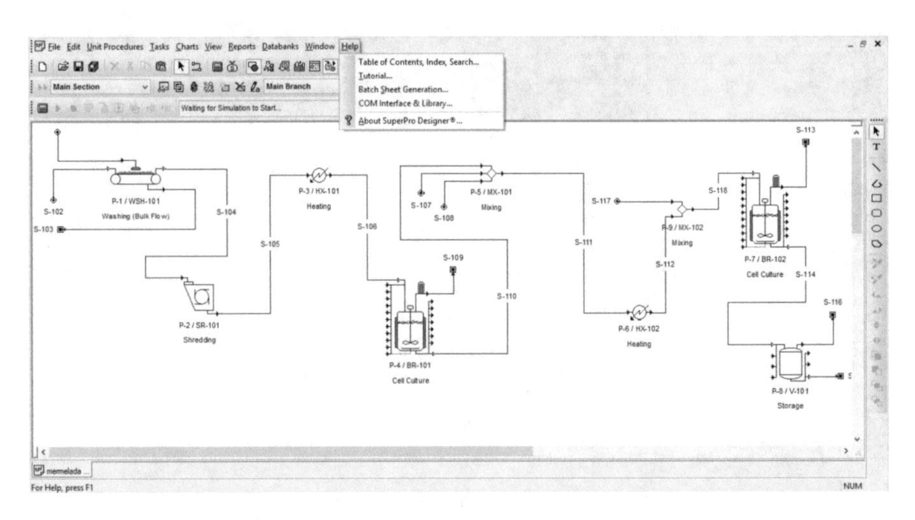

Fig. 4.22 Screenshot of COM interface in SuperPro Designer®

4.3.3 *Call Name of SuperPro Designer® Variables*

Once the path of the backup file of the process simulation has been declared, we proceed with the declaration of decision variables and response variables in the linker program. To do this, we need to open the file containing the simulation in SuperPro Designer® and go to Help > COM interface (Fig. 4.22).

Then, you will see the COM Library where we proceed to find the name of the variable that uses the COM interface and that corresponds to the process variable that we want to link (Fig. 4.23).

The user must look in the tree of the COM Library, the variable that wants to bind either as a decision variable or as a response variable. In order to find the desired

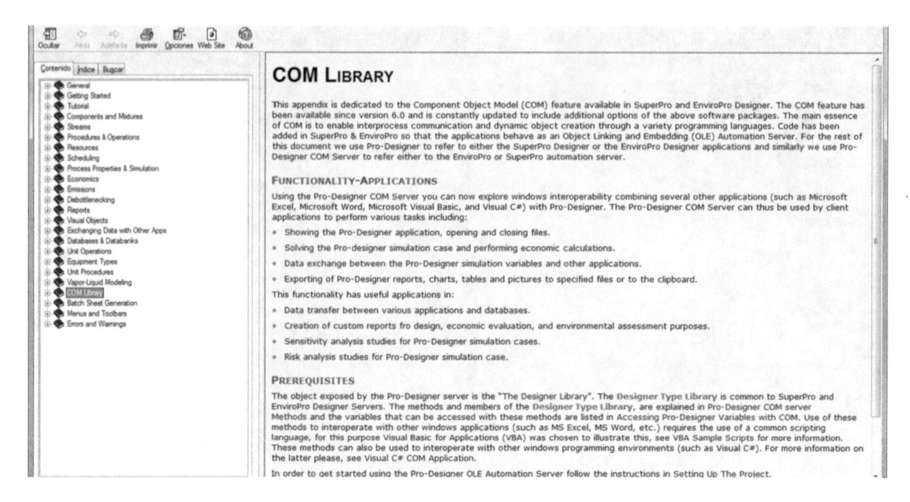

Fig. 4.23 Screenshot of variable explorer tree in Aspen Plus®

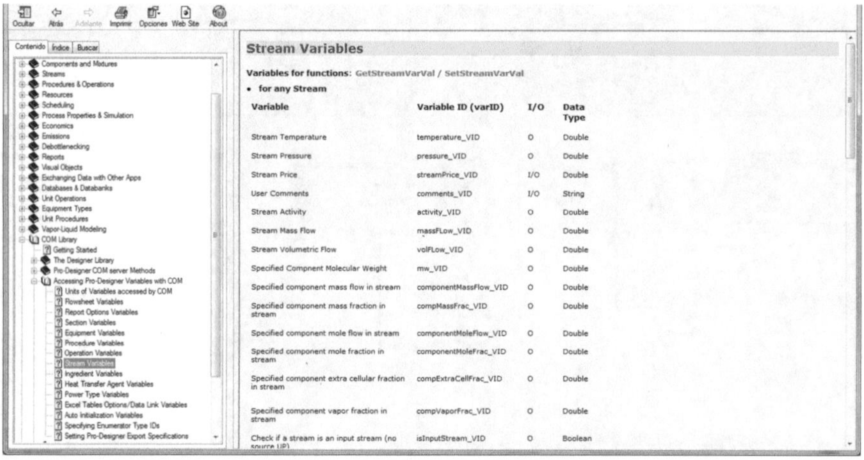

Stream Variables

Variables for functions: GetStreamVarVal / SetStreamVarVal

- for any Stream

Variable	Variable ID (varID)	I/O	Data Type
Stream Temperature	temperature_VID	O	Double
Stream Pressure	pressure_VID	O	Double
Stream Price	streamPrice_VID	I/O	Double
User Comments	comments_VID	I/O	String
Stream Activity	activity_VID	O	Double
Stream Mass Flow	massFLow_VID	O	Double
Stream Volumetric Flow	volFLow_VID	O	Double
Specified Compnent Molecular Weight	mw_VID	O	Double
Specified component mass flow in stream	componentMassFlow_VID	O	Double
Specified component mass fraction in stream	compMassFrac_VID	O	Double
Specified component mole flow in stream	componentMoleFlow_VID	O	Double
Specified component mole fraction in stream	componentMoleFrac_VID	O	Double
Specified component extra cellular fraction in stream	compExtraCellFrac_VID	O	Double
Specified component vapor fraction in stream	compVaporFrac_VID	O	Double
Check if a stream is an input stream (no source UP)	isInputStream_VID	O	Boolean

Fig. 4.24 Screenshot of COM Library tree in SuperPro Designer®

variable, a logical sequence of search is followed according to the equipment or stream to which this variable belongs. For example, if the user is looking for the variable "Feeding temperature" of an equipment, the path to be followed in the COM Library tree is as follows: COM Library > Accessing SuperPro Designer® Variable with COM > Stream Variables > Stream Temperature. Once the path has been followed to the desired variable, a series of variables appears on the right side of the COM Library. Since we are sure that it is the searched variable, we copy the contents of the value corresponding to the "Variable ID" column (Fig. 4.24).

Once you have copied the name of the variable with which the COM technology will make the call, proceed to open the link subroutine again, and look for the part in which the variables are assigned; pasted here is the variable name for the call (Fig. 4.25).

Fig. 4.25 Screenshot of linking subroutine where the variable name call must be pasted

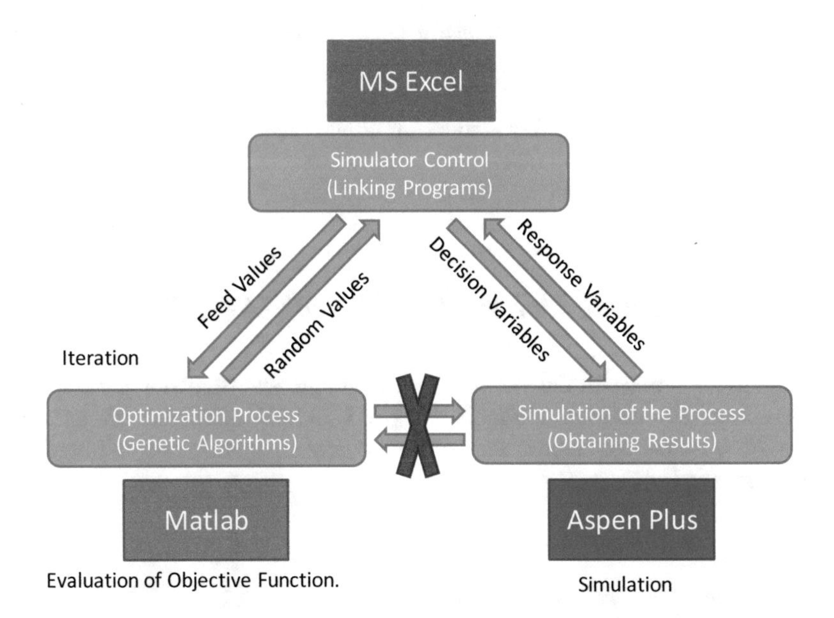

Fig. 4.26 Interface between Aspen Plus® and MATLAB® using MS Excel® as linking program

4.4 Link Between MS Excel® and MATLAB®

An interface between the process simulation software Aspen Plus® and MATLAB® can be established using MS Excel® as linking program (as can be seen in Fig. 4.26). Direct communication between the process simulation software (Aspen Plus®) and the program containing the stochastic optimization tools (MATLAB®) cannot be established; therefore, it is necessary to use a third program that works

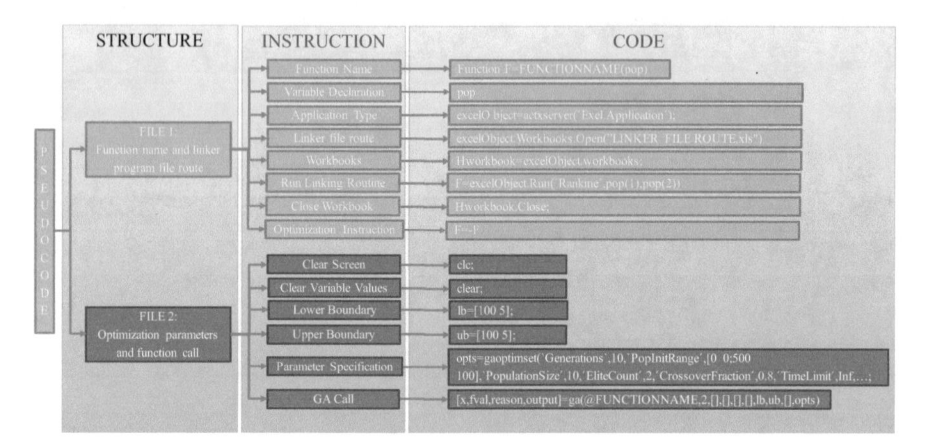

Fig. 4.27 Pseudo code for the link between MS Excel® and MATLAB®

as a linker. Due to the multiple advantages of the use of COM technology and because there are programs that easily allow their use, MS Excel® has been selected as a linker program.

For a detailed information about how to make this link, we recommend reviewing the tutorial video that is included as additional material for this book. You can access it using the following link:

http://extras.springer.com

4.4.1 Subroutine to Link MS Excel® and MATLAB®

To perform the successful link between a software process simulator and a program that has stochastic optimization tools using MS Excel® as a linker program, it is necessary to follow the steps described by the general syntaxes for the used code which is shown in Fig. 4.27.

4.4.2 Files Needed to Link MS Excel® and MATLAB®

Before you begin, you need four files to perform the link (shown in Fig. 4.28). The first one corresponds to the backup file of Aspen Plus® that corresponds to the data of the simulation processes previously elaborated. The second file will be from the MS Excel® linker program, which will include a routine that will call the variables of the simulator, placing input values and reading the response variables. The following two files will be from the program containing the MATLAB® stochastic optimization tool; one of them declares the used function by the

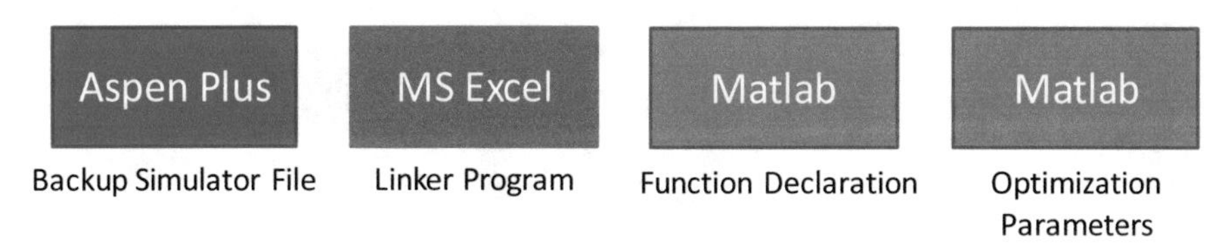

Fig. 4.28 Four files needed to start the linking Aspen Plus®, MATLAB®, and MS Excel®

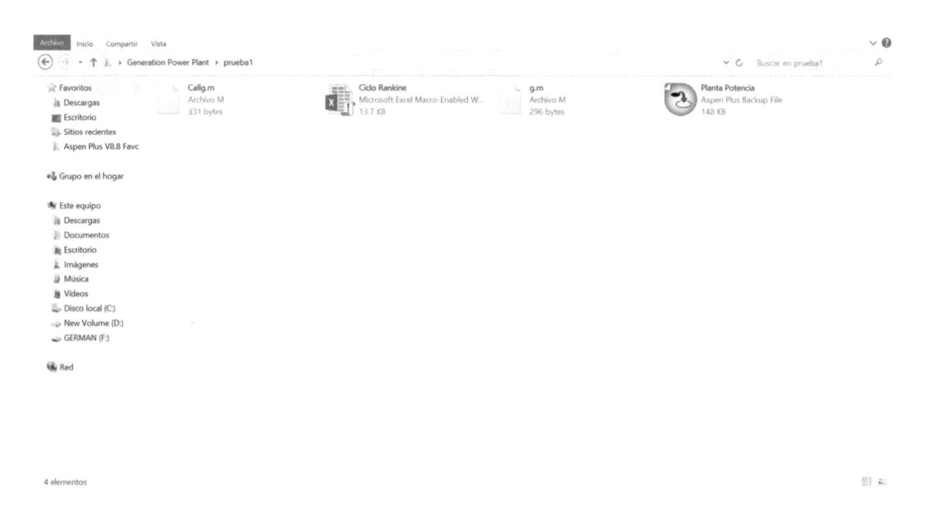

Fig. 4.29 Screenshot of windows explorer with the four files needed to start linking Aspen Plus®, MATLAB®, and MS Excel®

optimization, and the other file introduces the parameters that will be used in the genetic algorithm and serves to make the call of the declared function.

Figure 4.29 shows a screenshot of windows explorer with the four files needed to start linking Aspen Plus®, MATLAB®, and MS Excel®. The orders in which these files appears from left to right are optimization parameters, linker program, function declaration, and backup simulator file. It is very important to distinguish each of these files to avoid future confusions; the most practical and simple way to do this is through the characteristic symbol of each program; however, it can also be done by the extension of each file (.bkp for Aspen Plus®, .xls for MS Excel®, and .m for MATLAB®). It is also recommended to use clear names that the user can easily relate to the contents of each file.

Once there has been corroborated that you have the essential files for the link, we can begin with the explanation of the steps to follow, which are described next.

If the user does not have any files for the process simulation in Aspen Plus®, for the optimization algorithm in MATLAB®, and for the linker program in MS Excel®, the files shown in Fig. 4.29 can be found in the following link:

http://extras.springer.com

Fig. 4.30 Screenshot of windows explorer with the "Properties" option of the linker program

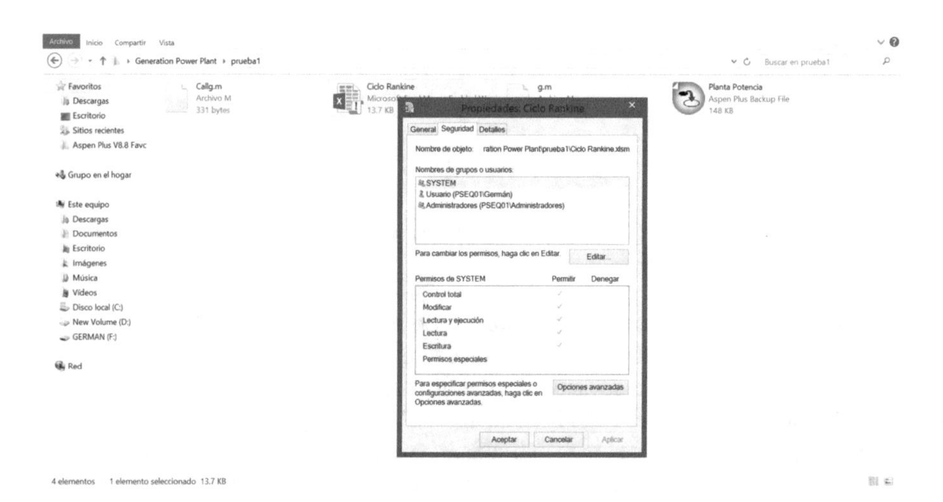

Fig. 4.31 Screenshot of windows explorer with the "Security" tab of the linker program where the "Object Name" can be seen

4.4.3 Object Name of the Linker Program File

In a similar way to the simulator file route declaration in the linker program, it is necessary to open the windows explorer in the folder containing the four files needed for linking. Then, to check the exact route of the simulation, select the Aspen Plus® backup file and right click on it, and the same menu will be displayed where we will select the last option "Properties" (Fig. 4.30).

As expected, a window will open where we select the "Security" tab and copy the whole path of the box "Object Name" (Fig. 4.31).

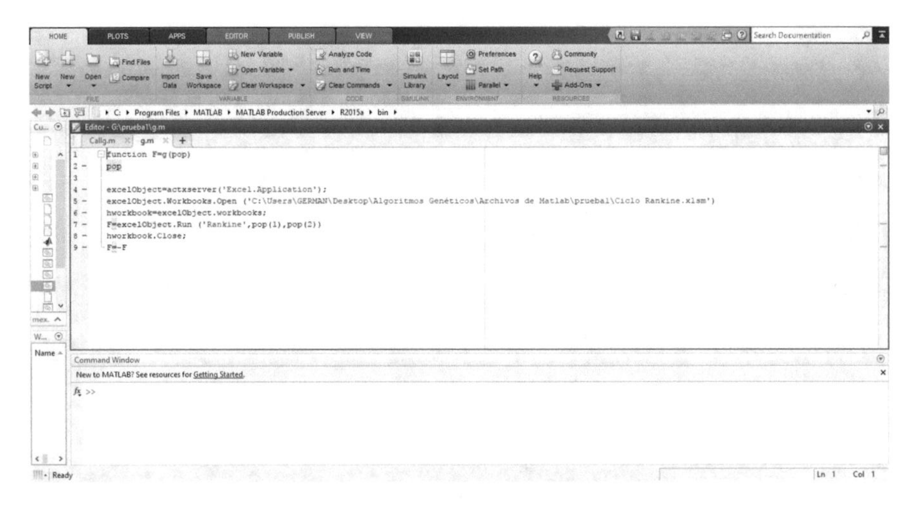

Fig. 4.32 Screenshot of function declaration file, where the linker program route must be pasted

Now, you need to open the file containing the function to be optimized in MATLAB®. In this file, it is necessary to look for the part in which the file of the linker program is declared as object, which must be the fifth line of the code. Once this instruction is located, we proceed to paste the file path of the linker program (Fig. 4.32).

4.4.4 *Specification for the Optimization Parameters in MATLAB®*

In this part, the simulator file route has been declared in the linker program, the variables have been declared in the linker program, and the linker program file route has been declared in the function file. What corresponds now is the specification of the variables related to the MATLAB® stochastic optimization tool; so the first thing to do is open the file containing the function call. In this file, we proceed to indicate the upper and lower search limits for the variables to be optimized. In addition, the optimization parameters must be specified according to the considerations that we have to get an adequate search (Fig. 4.33).

Each process has different parameters so it is impossible to standardize them, so the user must know the process very well to propose the ones that work best. For more information about each parameter used in genetic algorithms, we recommend that the reader review the previous chapter about process optimization and stochastic search algorithms.

After that, it only remains to run the optimization, which is done from this last MATLAB® file by clicking on the "Run and Time" option.

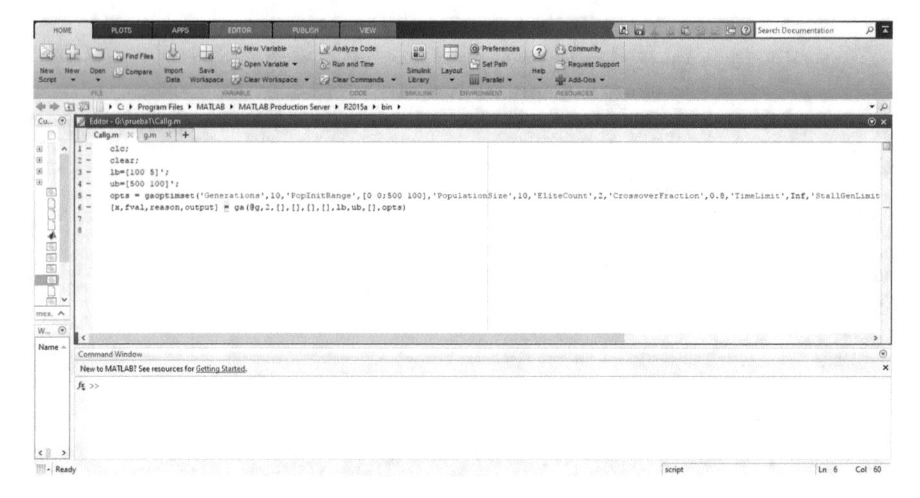

Fig. 4.33 Screenshot of optimization parameter file, where the function is called

4.5 Exercises

1. The following link shows the simulation process of a reactive distillation process, this corresponds to the process flowsheet in Aspen Plus® for the production of methyl tert-butyl ether (MTBE):http://extras.springer.com

   (a) Implement the link between Aspen Plus® and the I-MODE algorithm in MS Excel® specifying the path where it was saved as the simulation program file.
   (b) Declare the following decision variables in the main program user interface with their respective lower and upper limits: number of plates (5–8) and stage of feeding (3–5).
   (c) Run the optimization and find the optimal values for these variables.

2. Martinez-Gomez et al. (2017) reported an optimization-based approach for incorporating economic and safety considerations in the selection of reforming technology for the production of syngas from shale gas. Use the process flowsheet shown in Fig. 4.34 to implement the following:

   (a) Implement a link between Aspen Plus® and GA from MATLAB® using MS Excel® as a linker program, specifying the path where it was saved as the simulation program file.
   (b) Declare the following decision variables in the GA from MATLAB® with their respective lower and upper limits: temperature (1100–1200 K) and pressure (0.1–0.2 MPa).
   (c) Run the optimization and find the optimal values for these variables.

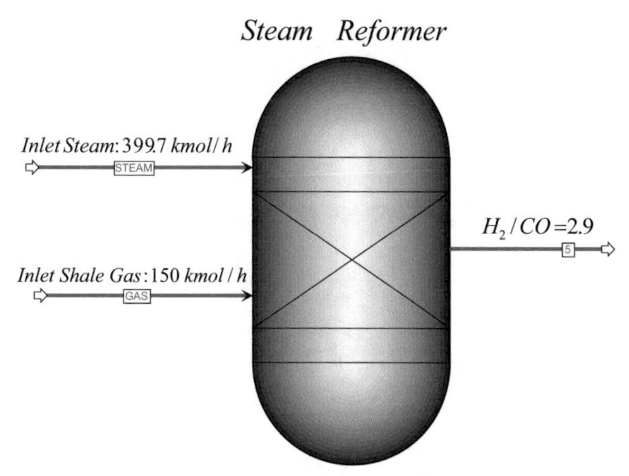

Fig. 4.34 Steam reformer for the production of syngas from shale gas

4.6 Nomenclature

COM	Component Object Module
MS	Microsoft®
MTBE	Methyl tert-butyl ether
VBA	Visual Basic for Applications

Chapter 5
Performance Evaluation

The performance evaluation is of vital importance to determine the effectiveness of both the search methods and the optimization parameters. Likewise, the performance evaluation allows us to evaluate if the selected decision variables have a considerable impact on the final result of the optimization and if the search range is adequate. This performance evaluation is achieved through one or several objective functions that the different search algorithms will try to satisfy at the same time as the restrictions specified by the user.

As it is mentioned above, the objective function is an equation in which is reflected the performance of the process that is being optimized; it is achieving its maximum or its minimum value by manipulating the variables of dissolution and considering the established search restrictions.

5.1 Objective Functions

It should be noted that there are many and varied ways to calculate an objective function; this will depend on what the user seeks. Next, some of the main objective functions commonly used in optimization will be described.

5.1.1 Net Present Value

A very common way of objective function is the net present value (NPV), which is an economic measure that considers the value of money over time. It is the present value of an investment's future net cash flow (= difference between the money coming in and going out) after the cost of the original investment has been subtracted (Cambridge dictionary).

J. M. Ponce-Ortega, L. G. Hernández-Pérez, *Optimization of Process Flowsheets through Metaheuristic Techniques*,
https://doi.org/10.1007/978-3-319-91722-1_5

5.1.2 *Profit*

Another important objective function is the profit. The profit is money that is earned in trade or business after paying the costs of producing and selling goods and services (Cambridge dictionary).

5.1.3 *Capital Cost*

The capital cost is the amount of money that a company must pay out in dividends to its shareholders and in interest on bonds and other loans (Cambridge dictionary).

5.2 Capital Cost Estimation Programs

There are many tools to achieve a good estimation of the capital cost. Some of which can be programmed directly by the user in different platforms. Likewise, there are some programs that are dedicated exclusively to the estimation of the capital cost and that allow obtaining good values from the requested data. Feng and Rangaiah (2011) made an evaluation of capital cost estimation programs in which five programs were compared using a set of case studies. The most popular programs for the capital cost estimation are CapCost, EconExpert, AspenTech Process Economic Analyzer (Aspen-PEA), detailed factorial method (DFP), and capital cost estimation program (CCEP).

5.2.1 *CapCost*

The CapCost is based on the module costing method, written in Visual Basic, and can be used for estimating preliminary process cost. Bare module cost (CBM) is defined as the sum of the direct and indirect expenses for purchasing and installing equipment; the total module cost (CTM) is defined as the sum of the bare module cost, contingency, and fee; and the grassroots plant cost (CGR) is defined as the sum of the total module cost and the auxiliary facilities costs. To estimate the bare module cost and purchase cost of equipment, Turton et al. (2009) proposed the following:

$$C_{\mathrm{BM}} = C_p^0 \times F_{\mathrm{BM}} = C_p^0 \left(B_1 + B_2 F_{\mathrm{M}} F_{\mathrm{P}}\right) \tag{5.1}$$

$$\log C_p^0 = K_1 + K_2 \log(S) + K_3 \left[\log(S)\right]^2 \quad (5.2)$$

where S represents a parameter for the equipment size or capacity. Values for the constants B_1 and B_2, equipment-specific constants K_1, K_2, and K_3, as well as correlations and plots for F_{BM}, F_M, F_P, and Co of different equipment can be found in the appendices in Turton et al. (2009).

5.2.2 *Detailed Factorial Program (DFP)*

The detailed factorial program (DFP) is based on the detailed factorial estimates method described in Sinnott and Towler (2009). For this program, the purchase cost, C_p^0, of the major equipment items is estimated using the following:

$$C_p^0 = a + bS^n \quad (5.3)$$

Cost constants a and b, available in Sinnott and Towler (2009) for different equipment items, are mainly for carbon-steel material.

5.2.3 *Capital Cost Estimation Program (CCEP)*

Capital cost estimation program (CCEP) uses cost correlations in Seider et al. (2010) for estimation of free-on-board purchase cost of equipment. The material factor and Guthrie's bare module factor are used thereafter to estimate the installed cost of that equipment. Seider et al. (2010) developed the purchase cost correlations for common process equipment, based on available literature sources and vendor data. A list of these cost correlations can be found in Seider et al. (2010), using CEPCI = 500. The purchase cost of the major equipment items is estimated using the following:

$$C_p^0 = e\left\{A_0 + A_1\left[\ln(S)\right] + A_2\left[\ln(S)\right]^2 + \cdots\right\} \quad (5.4)$$

Values of constants A_0, A_1, and A_2 for various equipment items can be found in Seider et al. (2010). CCEP and DFP were developed in Microsoft Excel and Visual Basic environments, by Wong (2010) and Huang (2010), respectively, as part of research projects supervised by the second author (these programs can be obtained from the authors).

5.2.4 *EconExpert*

EconExpert is a Web-based interactive software for capital cost estimation (Vasudevan and Agrawal 1999). Similar to CapCost, the equipment module costing method is used to calculate bare module cost and total module cost from the purchase cost of equipment. The purchase cost data and bare module factors used can be found in Ulrich and Vasudevan (2004). In this book, the cost data are expressed in graphical form, whereas in EconExpert, the plots are represented as polynomial equations for calculation of the purchase cost. Multiple regression is used to fit the data if the purchase cost is dependent on more than one variable. The cost data and correlations in EconExpert are for a CEPCI of 400 (Ulrich and Vasudevan 2004).

5.2.5 *AspenTech Process Economic Analyzer (Aspen-PEA)*

AspenTech Process Economic Analyzer (Aspen-PEA) is built on Aspen Icarus technology and is designed to generate both conceptual and detailed estimates (AspenTech 2009). It takes a unique approach, representing equipment by comprehensive design-based installation models. Aspen-PEA claims to contain time-proven, field-tested, industry-standard cost modeling and scheduling methods (AspenTech 2009).

5.3 Nomenclature

A_0, A_1, and A_2	Values of constants
B_1 and B_2	Values for constants
CBM	Bare module cost
CCEP	Capital cost estimation program
CGR	Grassroots plant cost
CTM	Total module cost
DFP	Detailed factorial method
F_{BM}, F_M, F_P, and Co	Correlations for different equipment
K_1, K_2, and K_3	Equipment-specific constants
PEA	Process Economic Analyzer
S	Parameter for the equipment size or capacity

Chapter 6
Optimization of Industrial Process 1

To illustrate the application of the method described in the previous chapter, the multi-objective optimization problem of the regenerative steam power plant with superheat and reheat shown in Fig. 6.1 is taken as an example. It consists of one stage of steam reheat and two closed feed water heaters with drains cascaded backward that operates at different pressure levels. Each feed water heater is a heat exchanger that receives steam bled from the turbine and feed water or high-pressure subcooled liquid water from the condenser. The water stream passes through successive steam-fed preheaters from the turbines and the condensation of which causes the heat to flow to the boiler feed stream to preheat. As the bled steam condenses in each feed water heater, it is passed through a pressure reducing valve to flow to a lower pressure region, such as either the next lower-pressure feed water heater or the condenser. In the condenser, cooling water provided by a wet-cooling tower removes the waste heat from the turbine exhaust steam at the lowest pressure level of the plant, leaving subcooled liquid water or condensate for reuse in the cycle. A pump is placed after the condenser to deliver water through the three-high-pressure closed feed water heaters to the boiler. The boiler generates high-pressure superheated steam from boiler feed water by combusting natural gas. Superheated high-pressure steam from the boiler is used to generate electric power in HP, IP, and LP turbines.

6.1 Problem Statement

In this example, it was addressed the simultaneous economic and environmental optimization of regenerative-reheat steam power plants for electric generation as the one illustrated in Fig. 6.1. Given are the plant configuration, temperature, pressure, and flow rate of the boiler output stream and the feed stream to the condenser, hot stream outlet temperature in the condenser, hot stream temperature decrease in

J. M. Ponce-Ortega, L. G. Hernández-Pérez, *Optimization of Process Flowsheets through Metaheuristic Techniques*,
https://doi.org/10.1007/978-3-319-91722-1_6

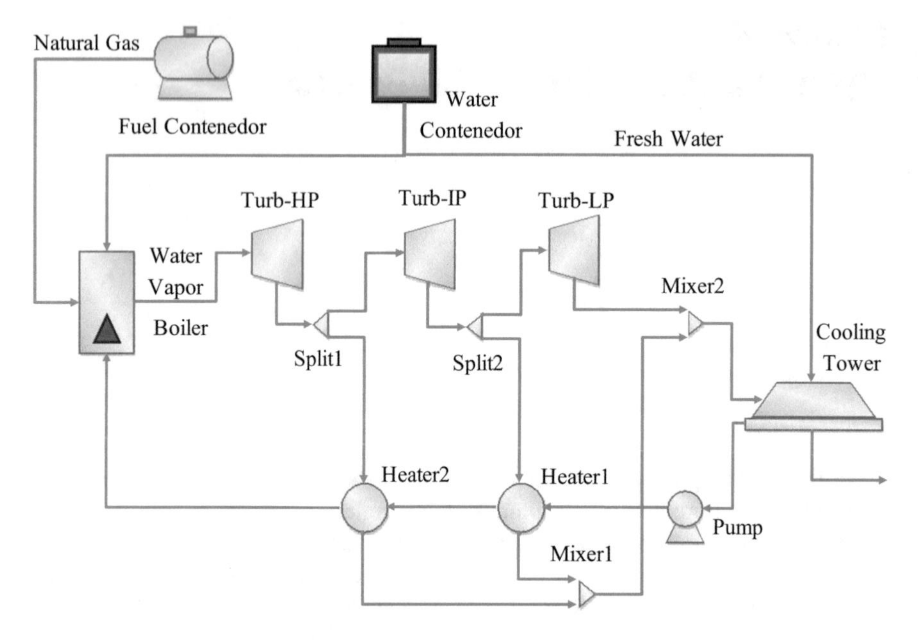

Fig. 6.1 Generation power plant based on regenerative Rankine cycle

both preheaters, pressure of the pump, temperature and pressure of the boiler, split fraction of both splitters, and pressure decrease in HP, IP, and LP turbines.

The solution of the problem is defined by a set of optimal designs called Pareto optimal set (i.e., the set of the best possible trade-offs between the considered objectives). Each of these solution alternatives achieves a unique combination of profit and environmental impact. For each solution of the Pareto set of the problem, the goal is to determine the optimal values of the temperature and pressure in the boiler, the pressure decrease in HP, IP, and LP turbines, the pressure in the pump, and the split fraction in both splitters as well as the optimal combination of energy sources that simultaneously maximizes the profit and minimizes the environmental impact of the plant (Gutiérrez-Arriaga et al. 2013).

6.2 Model Formulation

As can be seen in Fig. 6.1, simple electric power stations have configurations that comprise the following main components: a boiler; HP, IP, and LP turbines; a feed water pump; two feed water preheaters; and a cooling tower as condenser. A variety of steam power plant configurations can result from the different number, type, and connections of these components.

To facilitate the multi-objective optimization of such complex systems characterized by a large number of thermodynamic, economic, and environmental parameters, a simulation framework and a posterior optimization are proposed in this work.

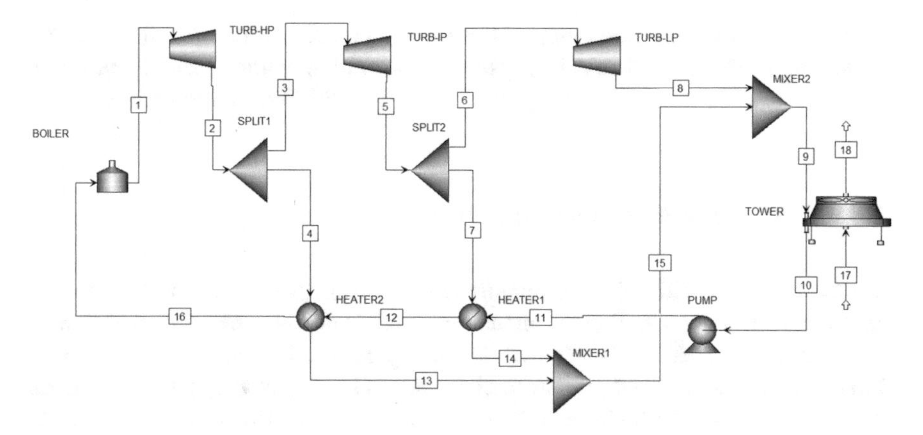

Fig. 6.2 Process flowsheet in Aspen Plus® for a simple steam power plant for electric generation based on regenerative Rankine cycle

In this section, we present the details of the proposed approach to tackle the problem described above taking as an example the regenerative steam power plant with superheat and reheat shown in Fig. 6.1.

6.2.1 Model Simulation Using the Aspen Plus® Software

The first step corresponds to the simulation of the process, i.e., to set the equipment and connections of streams that relate process units under the specific conditions of each one that allow to offer a representation of the process. For this purpose, Aspen Plus® was used, which is the market-leading chemical process simulation software used by the bulk, specialty, and biochemical industries for the design and operation (aspentech.com). The main advantages of this simulator consist in a large database of specific chemical compounds and unit operations. The process flowsheet in Aspen Plus® for a simple steam power plan for electric generation based on regenerative Rankine cycle is shown in Fig. 6.2.

Economic and environmental objective functions were defined through a mathematical formulation of the problem to be considered. The application of chemical and biochemical engineering simulators was not involved up today in the search of the optimum solutions. Due to this fact, the use of a multi-objective optimization algorithm is necessary which must be stochastic (e.g., genetic algorithm (GA) and differential evolution (DE)). A useful client-server application was developed in order to call Aspen Plus® simulator repetitively for various sets of input variables.

For the first simulation, the following values were used: a temperature of 580 °C, pressure of 38 atm, and total flow of 1000 ton/day for the boiler output stream. The hot stream outlet temperature in the condenser is equal to 100 °C, hot stream temperature decrease of 10 °C in the first preheater and 100 °C in the second one, pressure of the pump of 40 atm, temperature of 600 °C, and pressure of 40 atm in the boiler. The split fraction is 0.8 in both splitters, and the pressure decreases are 20, 10, and 5 atm in the HP, LP, and LP turbines, respectively.

6.2.2 Mathematical Formulation

In this step, the multi-objective optimization of steam power plants contains two objective functions including the annual gross profit and the environmental impact that must be satisfied simultaneously. For this purpose, the values of the response variables were used to calculate the performance of the objective functions using the following equations taken from Gutiérrez-Arriaga et al. (2015). The main benefit of using biofuels (i.e., natural gas) as energy sources in steam power plants is the reduction of the net CO_2 emissions (i.e., overall environmental impact). However, a lower environmental impact is associated with a lower plant annual gross profit. This poses a challenging multi-objective optimization problem of steam power plants where the overall environmental impact needs to be minimized while maximizing the system annual gross profit. The total income is calculated with the negative of electric energy produced by HP, IP, and LP turbines (WT) in kW and the electric power price of \$0.1039/kWh. The operating time (t_{OP}) was set to an average of 24 h for 360 days.

First, it is necessary to calculate the saturation temperature in the boiler (T_{sb}), which is calculated in °C using Eq. (6.1) starting with the value of the boiler pressure (P_b) expressed in atm:

$$T_{sb} = 13.8 P_b^{0.2264} \tag{6.1}$$

For calculating the bulb temperature of the boiler (T_{sh}) in °C, we used Eq. (6.2) introducing the boiler operating temperature (T_b) in °C:

$$T_{sh} = T_{sb} + T_b \tag{6.2}$$

Also, two dimensionless factors are necessary for calculating the capital cost of the boiler, the boiler superheat factor (N_t), and the cost factor in the boiler pressure (N_p), and they are calculated through Eqs. (6.3) and (6.4), respectively:

$$N_t = 1.5 \times 10^{-6} T_{sh}^2 + 1.13 \times 10^{-3} T_{sh} + 1 \tag{6.3}$$

$$N_p = 7 \times 10^{-4} P_b + 1 \tag{6.4}$$

The value of capital cost of the boiler (CB) is obtained by Eq. (6.5) using the dimensionless factors and the net heat required in the boiler operation ($Q_{netboiler}$) in kW:

$$\mathrm{CB} = \frac{3 \cdot N_{\mathrm{t}} \cdot N_{\mathrm{p}} \cdot Q_{\mathrm{netboiler}}^{0.77}}{3412.14} \tag{6.5}$$

while the cost of the pump (CP) is calculated through Eq. (6.6) using the value of the work done on the pump (WP) in kW:

$$\mathrm{CP} = 475.3 + 34.95 \cdot \mathrm{WP} - 0.0301 \cdot \mathrm{WP}^2 \tag{6.6}$$

Another cost is the one associated with the turbine (CT), which is found for the use of Eq. (6.7) starting with the electric energy produced by HP, IP, and LP turbines (WT) in kW:

$$\mathrm{CT} = 2.237 \cdot \mathrm{WT}^{0.41} \tag{6.7}$$

The cost of the cooling tower (CC), given in Eq. (6.8), is calculated using the heat removed from the cooling tower (Q_{c}) in kW:

$$\mathrm{CC} = 43 \cdot Q_{\mathrm{c}}^{0.68} \tag{6.8}$$

The cost for operating the pump (COPP) is the electrical energy consumption in the operation of this equipment in kW, which is calculated by Eq. (6.9) starting with the work done in the pump (WP):

$$\mathrm{COPP} = \frac{\mathrm{WP} \cdot 0.1039 \cdot 8640}{0.6} \tag{6.9}$$

The operating costs of the boiler (COPB) and cooling tower (COPC) are taken from the costs of the utilities used; these are variables that Aspen Plus® provides specifying the type of utility and the unit cost of everyone: for the boiler, it is a natural gas with a unit cost of 0.8552 \$/kg, and for the cooling tower, water was used as a cooling utility with a unit cost of 5.28×10^{-4} \$/kg. The capital cost factor (CCF) is taken into account for thermoelectric plants with a value of 0.1.

6.2.3 Definition of the Objective Functions

The gross annual profit (to be maximized) and the environmental impact (to be minimized) of steam power plants are taken as the two objectives to be simultaneously optimized. Next, we present the equations used to calculate these objective functions.

6.2.4 *Economic Objective Function*

The economic objective function consists in the maximization of the gross annual profit, which represents the difference between the total income and the total annual cost of the steam power plant. The performance of the economic objective function is calculated repetitively using the presented equations starting with the response variables obtained by the simulation software. The economic objective function is expressed in Eq. (6.10):

$$\begin{aligned}\text{Net Profit} = & \left(-\text{WT}\cdot\text{PkWh}\cdot t_{\text{OP}}\right)-\left(\text{CB}+\text{CP}+\text{CT}+\text{CC}\right)\cdot\text{CCF}\\ & -\left(\text{COPP}+\text{COPB}+\text{COPC}\right)\end{aligned} \tag{6.10}$$

6.2.5 *Environmental Objective Function*

In this study, the environmental objective function is to minimize the entire CO_2 emissions associated with electricity generation in power plants that use natural gas as primary energy source.

Aspen Plus® can calculate CO_2 emissions using US-EPA-Rule-E9-5711 as CO_2 emission factor data source with a value of $2.3e^{-07}$ kg/cal for natural gas. We assume a CO_2 energy source efficiency factor of 0.85, and starting with the needed heat in the boiler, which is a response variable calculated by Aspen Plus® after running the simulation of the power plant, we can calculate the total CO_2 emission.

6.3 Stochastic Optimization Algorithm Used

The multi-objective optimization hybrid method, namely, improved multi-objective differential evolution (I-MODE) developed by Sharma and Rangaiah (2013), is used as stochastic algorithm for the optimization of the process in this example. This improved multi-objective differential evolution algorithm works with a termination criterion using the non-dominated solutions obtained as the search process.

There were selected eight decision variables and introducing a value for the lower and upper boundary. The values of the selected decision variables for the lower and upper bounds, respectively, are 590 and 610 °C for operation temperature in the boiler, 38 and 42 atm for the pressure in the boiler, 18 and 22 atm for the pressure decrease in the HP turbine, 8 and 12 atm for the pressure decrease in the IP turbine, 4 and 6 atm for the pressure decrease in the LP turbine, 38 and 42 atm for the pressure in the pump, and 0.7 and 0.9 for the split fraction in both splitters. All decision variables were selected as continuous variables and the initial value for each was the half between the minimum and the maximum possible value. The

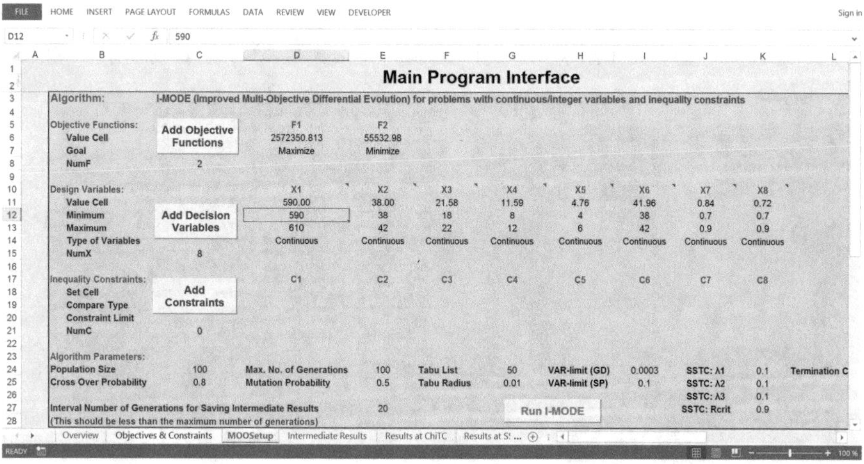

Fig. 6.3 MS Excel® sheet were main program user interface of the I-MODE (case 1)

optimization was developed without any inequality constraint. These values of the decision variables are introduced into the Main Program User Interface of the I-MODE as shown in Fig. 6.3.

For the optimization process, in this case study, the values for the parameters associated with the used I-MODE algorithm are the following: population size (NP) of 100 individuals, generation number (GenMax) of 100, taboo list size (TLS) of 50 individuals, taboo radius (TR) of 0.01, crossover fractions (Cr) of 0.8, and mutation fractions (F) of 0.5. These values of the parameters associated with the used of the algorithm are also introduced into the main program user interface of the I-MODE as shown in Fig. 6.3.

6.4 Link Between the Process Simulator and Optimization Algorithm

For the adequate link between the process simulator software (Aspen Plus for this example) and the stochastic optimization algorithm (the I-MODE in this case), it is necessary to follow the methodology mentioned in previous chapters. It is recommendable to add two more MS Excel® sheets, the first one for the decision variable values that will be sent to the simulator (Fig. 6.4) and the second one for the response variable values that will be received from the simulator (Fig. 6.5).

As can be seen, the additional equations of the mathematical formulation must be introduced in the MS Excel® sheet shown in Fig. 6.5. After that, the appropriated internal link between the decision variables, response variables, and objective functions must be established. Then, run the I-MODE since main program user interface.

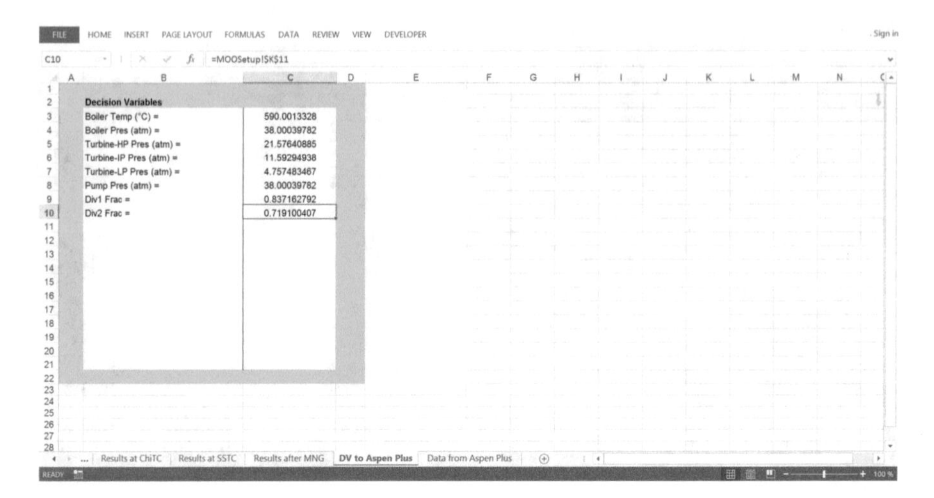

Decision Variables	
Boiler Temp (°C) =	590.0013328
Boiler Pres (atm) =	38.00039782
Turbine-HP Pres (atm) =	21.57640885
Turbine-IP Pres (atm) =	11.59294938
Turbine-LP Pres (atm) =	4.757483467
Pump Pres (atm) =	38.00039782
Div1 Frac =	0.837162792
Div2 Frac =	0.719100407

Fig. 6.4 MS Excel® sheet were decision variable values will be sent to the process simulator (case 1)

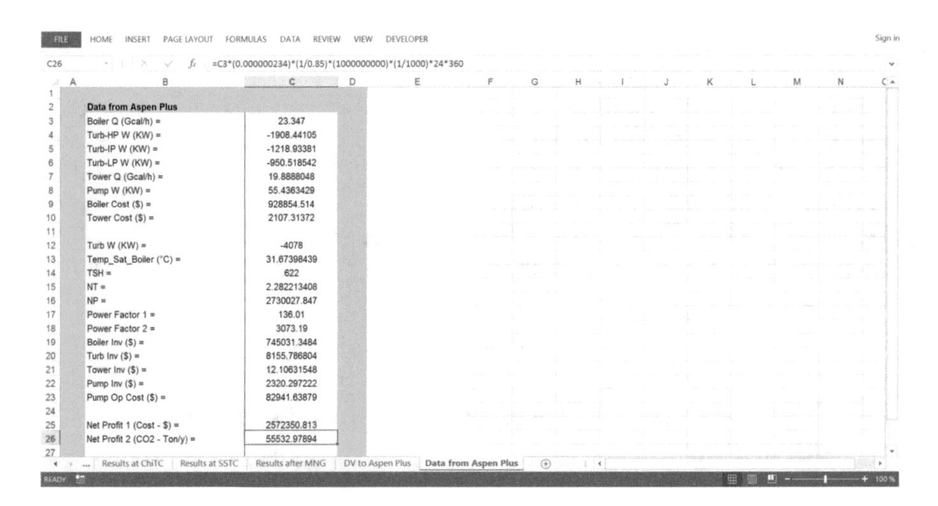

Data from Aspen Plus	
Boiler Q (Gcal/h) =	23.347
Turb-HP W (KW) =	-1908.44105
Turb-IP W (KW) =	-1218.93381
Turb-LP W (KW) =	-950.518542
Tower Q (Gcal/h) =	19.8888048
Pump W (KW) =	55.4363429
Boiler Cost ($) =	928854.514
Tower Cost ($) =	2107.31372
Turb W (KW) =	-4078
Temp_Sat_Boiler (°C) =	31.67398439
TSH =	622
NT =	2.282213408
NP =	2730027.847
Power Factor 1 =	136.01
Power Factor 2 =	3073.19
Boiler Inv ($) =	745031.3484
Turb Inv ($) =	8155.786804
Tower Inv ($) =	12.10631548
Pump Inv ($) =	2320.297222
Pump Op Cost ($) =	82941.63879
Net Profit 1 (Cost - $) =	2572350.813
Net Profit 2 (CO2 - Ton/y) =	55532.97894

Fig. 6.5 MS Excel® sheet were response variable values will be received from the simulator (case 1)

6.5 Results

This section presents the results of the multi-objective optimization method applied to the case study described in this chapter. All the runs were obtained from an Intel(R) Core TM i7-4700MQ CPU at 2.4 GHz, 32 GB computer; the computing time required to obtain the Pareto optimal solutions varied from 10 to 15 min.

The proposed strategy yields the Pareto sets shown in Figs. 6.6, 6.7, and 6.8, which show the optimal solution generated according to the stochastic procedure of

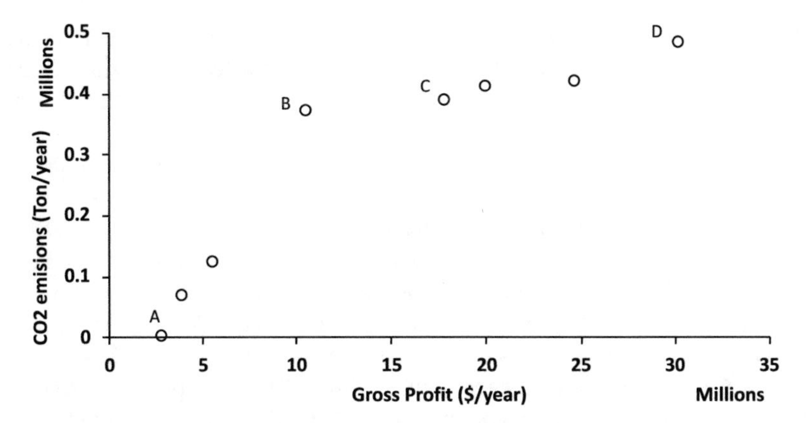

Fig. 6.6 Graphic of the results at the Chi-squared termination criterion (ChiTC)

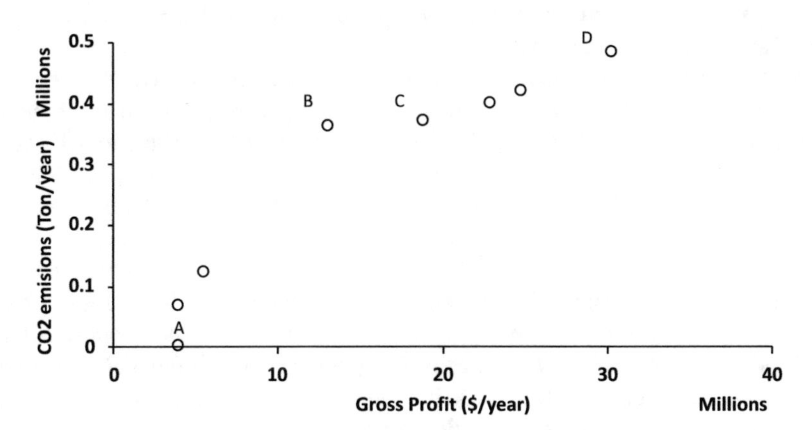

Fig. 6.7 Graphic of the results at the steady-state termination criterion (SSTC)

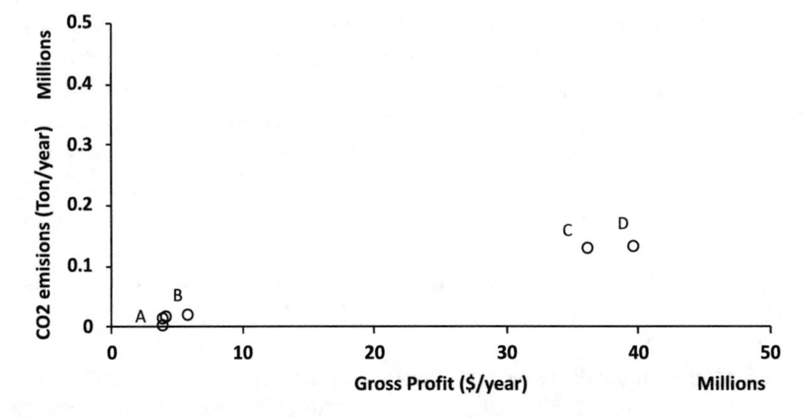

Fig. 6.8 Graphic of the results of the last generation

this method. The three different presented plots depend on the termination criteria. The shown Pareto plots were obtained starting with the selected decision variables, their values for the lower and upper bounds, and the values for the parameters associated with the used I-MODE algorithm presented in Chap. 3.

The results for the Chi-squared termination criterion (ChiTC) are shown in Fig. 6.6, which converges in 37 generation. In this plot, four important points can be seen (A, B, C, and D). In point A is shown the minimum value for CO_2 emissions (0 ton/year), but this point has a gross profit of 2,922,390 $/year, which is low. In point B and point C, there can be seen acceptable values for both objective functions (CO_2 emissions of 369,344 ton/year with a gross profit of 10,624,510 $/year for point B and CO_2 emissions of 385,858 ton/year with a gross profit of 17,891,507 $/year for point C). And point D shows the maximum value for the gross profit (30,330,600 $/year), but this point has the maximum value for CO_2 emissions too (483,497 ton/year). After the analysis of the graphic shown in Fig. 6.6, it was concluded that the best point is C because it offers a better gross profit than point B with a minimum increment in the CO_2 emissions.

The results for the steady-state termination criterion (SSTC) are shown in Fig. 6.7, which converges in 53 generations. In this graphic, four important points can be seen (A, B, C, and D). In point A is shown the minimum value for CO_2 emissions (0 ton/year), but this point has a gross profit of 3,975,780 $/year, which is low. In point B and point C, there can be seen acceptable values for both objective functions (CO_2 emissions of 361,757 ton/year with a gross profit of 13,161,987 $/year for point B and CO_2 emissions of 370,516 ton/year with a gross profit of 18,896,151 $/year for point C). And point D shows the maximum value for the gross profit (30,330,600 $/year), but this point is the maximum value for CO_2 emissions too (483,497 ton/year). After the analysis of the graphic shown in Fig. 6.7, it was concluded that the best point is C because it offers a better gross profit than point B with a minimum increment in the CO_2 emissions.

And the results for the last generation are shown in Fig. 6.8. In this graphic, four important points can be seen (A, B, C, and D). In point A is shown the minimum value for CO_2 emissions (0 ton/year), but this point (just as in the graphic shown in Fig. 6.7) has a gross profit of 3,975,780 $/year, which is low. Point B shows values not much different of point A (CO_2 emissions of 18,597 ton/year with a gross profit of 5,868,030 $/year). In point C and point D, there can be seen acceptable values for both objective functions (CO_2 emissions of 126,794 ton/year with a gross profit of 30,194,163 $/year for point C and CO_2 emissions of 130,249 ton/year with a gross profit of 39,687,071 $/year for point D). Point D shows the maximum value for the gross profit and the maximum value for CO_2 emissions too, but this point is not much different than point C in the value of CO_2 emissions, and it offers a considerable increment in the value of the gross profit. Based on this, it was concluded that the best point is D.

The I-MODE algorithm gives the optimal values for all the decision variables. The optimal values of the selected decision variables after running the optimization are the following: 590 °C for operation temperature in the boiler, 38.00 atm for the pressure in the boiler, 21.58 atm for the pressure decrease in the HP turbine,

11.59 atm for the pressure decrease in the IP turbine, 4.76 atm for the pressure decrease in the LP turbine, 41.96 atm for the pressure in the pump, and 0.84 and 0.72 for the split fraction in the first and second splitters, respectively.

The optimal value of the economic objective function, which consists in the maximization of the annual gross profit, is $2,572,350/year. The optimal value of the environmental objective function, which consists in the minimization of the entire CO_2 emissions associated with electricity generation in power plants that use natural gas as primary energy source, is 55,532 ton/year.

6.6 Exercises

To download the example of the Generation Power Plant, please click on the following link:

http://extras.springer.com

1. Use the process flowsheet of a regenerative Rankine cycle made in Aspen Plus just as shown in this chapter (Fig. 6.2).
2. Do the following (remember run the simulation after change of any specification):
   (a) Change the operation specification in the boiler, temperature of 590 °C and pressure of 18 atm (the discharge pressure of the pump must be of 18 atm too). What happen with the value of the generated work in the turbines? Why?
   (b) Change the total flow rate of the stream number 1, with a value of 2000 ton/day. What happen with the value of the generated work in the turbines? Why?
3. Use the main program user interface of the I-MODE algorithm shown in Fig. 6.3.
4. Implement the following:
   (a) Change the lower and upper bounds of the decision variables, 580 and 620 °C for the operation temperature in the boiler, 38 and 42 atm for the pressure in the boiler, 16 and 24 atm for the pressure decrease in the HP turbine, 6 and 14 atm for the pressure decrease in the IP turbine, 2 and 8 atm for the pressure decrease in the LP turbine, 36 and 44 atm for the pressure in the pump, and 0.65 and 0.95 for the split fractions in both splitters.
   (b) For the optimization process, in this case study, the values for the parameters associated with the used I-MODE algorithm are the following: population size (NP) of 1000 individuals, generation number (GenMax) of 1000, taboo list size (TLS) of 50 individuals, taboo radius (TR) of 0.01, crossover fractions (Cr) of 0.9, and mutation fractions (F) of 0.6.
   (c) Apply the same methodology to the conventional Rankine cycle, choose four decision variables, and propose different objective functions. Explain the obtained results.

6.7 Nomenclature

CB	Cost of the boiler
CC	Cost of the cooling tower
CP	Cost of the pump
CT	Cost of the turbine
COPB	Cost of operation of the boiler
COPC	Operation cost of the cooling tower
COPP	Cost of operation of the pump
COPT	Turbine operating cost
FCC	Capital cost factor
Net profit	Net profit
N_p	Cost factor in the boiler pressure
N_t	Overheating factor in the boiler
P_b	Boiler outlet pressure
P_c	Pressure of the cooling tower
P_p	Discharge pressure of the pump
P_t	Turbine output pressure
Q_b	Heat produced in the boiler
Q_c	Heat removed from the cooling tower
T_b	Temperature of the boiler
T_c	Temperature of the cooling tower
T_p	Temperature in the pump
T_t	Turbine output temperature
T_{sb}	Saturation temperature in the boiler
T_{sh}	Wet bulb temperature
W_n	Electric energy produced in the turbine
W_p	Work required by the pump

Chapter 7
Optimization of Industrial Process 2

This example analyzes the production of biodiesel from degummed soybean oil. The included SuperPro Designer® model is a slightly modified version of a process model developed by Haas et al. (2006).

There has been intense investigation on the development of fuel producing processes that are based on the use of renewable agricultural materials as feedstock. This activity is driven by the quest of national fuel self-reliance as well as reducing emissions of particulates, hydrocarbons, and carbon monoxide. Most of efforts have been concentrated on bioethanol and biodiesel. Biodiesel consists of the simple alkyl esters of the fatty acids found in agricultural acylglycerol-based fats and oils. It has been shown to give engine performance similar to that of conventional fuels.

Biodiesel can be produced from any material that contains fatty acids. Thus, various vegetable fats and oils or animal fats can be used as feedstocks for the biodiesel process. The choice depends on local availability, cost, and government regulations.

Here are the three most dominant ways of biodiesel production:

- Base-catalyzed transesterification of the oil
- Direct acid-catalyzed transesterification of the oil
- Conversion of the oil to its fatty acids and then to biodiesel

Most of the biodiesel produced today is done with the base-catalyzed reaction for several reasons:

- It requires low temperature and pressure.
- It yields high conversion (98%) with minimal side reactions and reaction time.
- It is a direct conversion to biodiesel with no intermediate compounds.
- No need for exotic materials of construction.

The chemical reaction for base-catalyzed biodiesel production is depicted below (Fig. 7.1).

J. M. Ponce-Ortega, L. G. Hernández-Pérez, *Optimization of Process Flowsheets through Metaheuristic Techniques*,
https://doi.org/10.1007/978-3-319-91722-1_7

The Biodiesel Reaction

CH_2OCOR'''				CH_2OH		$R'''COOR$
|			Catalyst	|		
CH_2OCOR''	+	3 ROH	------>	CH_2OH	+	$R''COOR$
|				|		
CH_2OCOR'				CH_2OH		$R'COOR$
100 pounds Oil or Fat		10 pounds Alcohol (3)		10 pounds Glycerin		100 pounds Biodiesel (3)

Fig. 7.1 Biodiesel formation reaction

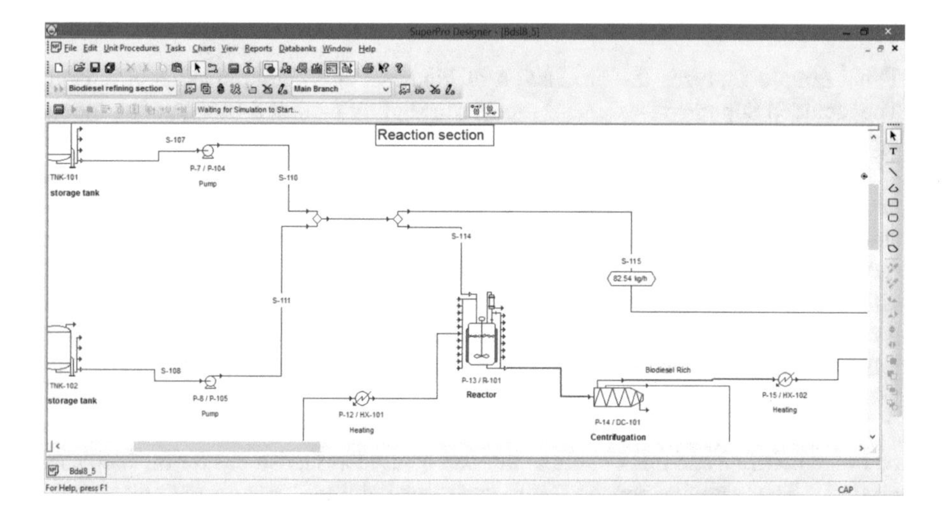

Fig. 7.2 Process flowsheet in SuperPro Designer® for a biodiesel production plant from degummed soybean oil

7.1 Problem Statement

One hundred pounds of fat or oil (such as soybean oil) are reacted with 10 pounds of a short-chain alcohol in the presence of a catalyst to produce 10 pounds of glycerol and 100 pounds of biodiesel. It is a reversible reaction so the short-chain alcohol, signified by ROH (usually methanol, but sometimes ethanol), is charged in excess to ensure quick conversion. The catalyst is usually sodium or potassium hydroxide that has already been mixed with methanol. R', R'', and R''' indicate the fatty acid chains associated with the oil or fat which are largely palmitic, stearic, oleic, and linoleic acids for naturally occurring oils and fats. Part of the process described before is shown in Fig. 7.2.

7.2 Model Formulation

For simplification purposes, the process has been split into three sections: the reaction (blue icons), the biodiesel refining (black icons), and the glycerol purification (green icons) section (Fig. 7.2). A section in SuperPro Designer® is simply a set of unit procedures (processing steps).

Reaction Section
The reaction section consists of:

- The raw material storage tanks for the methanol (TNK-101), the catalyst (TNK-102), and the soybean oil (TNK-103)
- The two reactors (R-101 and R-102)
- A decanter centrifugal separator (DC-101)

The soybean oil is directly fed to the reactor (R-101). Methanol and the catalyst are mixed, and 90% of the mixture is fed to the first reactor. The rest (10%) is fed to the second reactor. According to the reaction mentioned in Fig. 7.1, methanol reacts with soybean oil and yields biodiesel and glycerol. Product is removed at a rate equal to the rate of charging the reactants and catalyst. The average residence time of materials in the reaction is 1 h. Glycerol, a co-product of the acylglycerol transesterification, separates from the oil phase as the reaction proceeds. The reaction extent is approximately 90%. The material is then fed to a centrifugal separator (DC-101) where the biodiesel and the soybean oil that have not reacted are separated from the glycerol-rich co-product phase. The latter is sent to the glycerol recovery unit.

The biodiesel stream, which also contains unreacted methanol, soybean oil, and catalyst, is fed into a second stirred tank reactor (R-102) along with the addition of the methanol-catalyst stream from the splitter (FSP-101). The reaction conditions are the same. The reaction extent in the second reactor is 90% which yields a combined conversion efficiency of 99%.

Again the mixture of methyl esters (biodiesel), glycerol, unreacted substrates, and catalyst exiting the second reactor is fed to another centrifugal separator (DC-102).

Biodiesel Refining Section
This section consists of:

- Two continuous centrifugal separators (DC-102 and DC-104)
- A mixing vessel (V-102)
- A vacuum dryer system (V-104 and GBX-101)
- The biodiesel storage tank (TNK-104)

The crude biodiesel stream is washed with acidified water at a pH of 4.5 in a mixing tank (V-102) to neutralize the catalyst and turn any soap into free fatty acids. The material is then fed to a continuous centrifugal separator (DC-104) to separate the biodiesel from the aqueous phase, which is fed to the glycerol recovery section. The crude biodiesel product must contain a maximum of 0.050% w/w water. This is achieved by using a vacuum dryer system (V-104 and GBX-101). It lowers the water content from 2.3% to app. 0.04%.

Glycerol Purification Section

This section consists of:

- Two mixing vessels (V-101 and V-103)
- A centrifugal separator (DC-103)
- Two distillation columns (C-101 and C-102)
- Two storage tanks (TNK-105 and TNK-106)

The produced glycerol during the transesterification process requires purification before it can be sold. The equipment is sized to remove methanol, the fatty acids, and most of the product to yield an 80% pure glycerol which is then sold to industrial glycerol refiners at a price of $0.33/kg.

Both glycerol streams (S-119 and S-132) and fatty acid contaminants (S-137) exiting the reactors are pooled and treated with acid (HCl) in V-101 to convert soaps into free fatty acids, which are subsequently removed by centrifugation (DC-103). The fatty acid stream is destined to disposal.

The glycerol stream is then neutralized with caustic soda (in V-103). The methanol contained in the glycerol stream is recovered by distillation (C-101) and recycled back to the first reactor (R-101). Finally, the glycerol stream is concentrated to reach 80% purity by another distillation step (C-102) that removes the water, which is recycled back to the mixing vessel V-102.

7.2.1 *Model Simulation Using the SuperPro Designer® Software*

This example analyzes the production of 33,635 metric tons (MT) per year of biodiesel using crude degummed soybean oil.

The following SuperPro Designer® flowsheet file has been included with this example:

- Bdsl8_0.spf

This flowsheet shows the base case for the process. The Bdsl8_0.spf file was used to produce the tables and graphs in the rest of this chapter.

Below is a brief description of the basic features of the biodiesel process (file Bdsl8_0.spf). All files for this example can be found in the "C:\Program Files\Intelligen\SuperPro Designer \ EXAMPLES\BioDiesl" directory.

7.2.2 *Definition of the Objective Functions*

The gross annual profit (to be maximized) and the environmental impact (to be minimized) of steam power plants are taken as the two objectives to be simultaneously optimized. Next, we present the equations used to calculate these objective functions.

7.2.3 *Economic Objective Function*

The economic objective function consists in the maximization of the gross annual profit, which represents the difference between the total income and the total annual cost of the biodiesel production plant.

7.2.4 *Environmental Objective Function*

In this study, the environmental objective function is to minimize the entire CO_2 emissions associated with heating utilities of the biodiesel production plant.

7.3 Stochastic Optimization Algorithm Used

The multi-objective optimization hybrid method, namely, improved multi-objective differential evolution (I-MODE) developed by Sharma and Rangaiah (2013), is used as stochastic algorithm for the optimization of the process in this example.

There were selected four decision variables and introducing a value for the lower and upper boundary. The values of the selected decision variables for the lower and upper bounds, respectively, are 60 and 70 psi for operation pressure in the stream 110, 60 and 70 psi for operation pressure in the stream 111, 55 and 65 °C for operation temperature in the stream 116, and 60 and 70 psi for operation pressure in the stream 116. These values of the decision variables are introduced into the main program user interface of the I-MODE as shown in Fig. 7.3.

For the optimization process, in this case study, the values for the parameters associated with the used I-MODE algorithm are the following: population size (NP) of 100 individuals, generations number (GenMax) of 100, taboo list size (TLS) of 50 individuals, taboo radius (TR) of 0.01, crossover fractions (Cr) of 0.8, and mutation fractions (F) of 0.5. These values of the parameters associated with the used of the algorithm are also introduced into the main program user interface of the I-MODE as shown in Fig. 7.3.

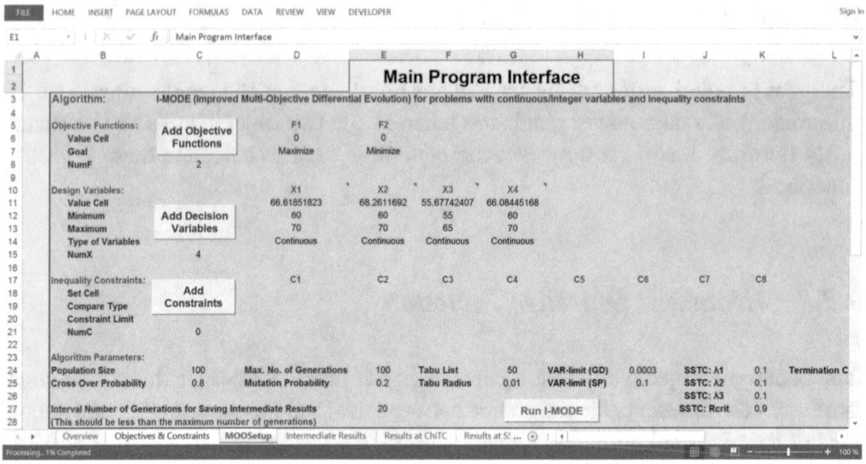

Fig. 7.3 Excel® sheet for the main program user interface of the I-MODE (case 2)

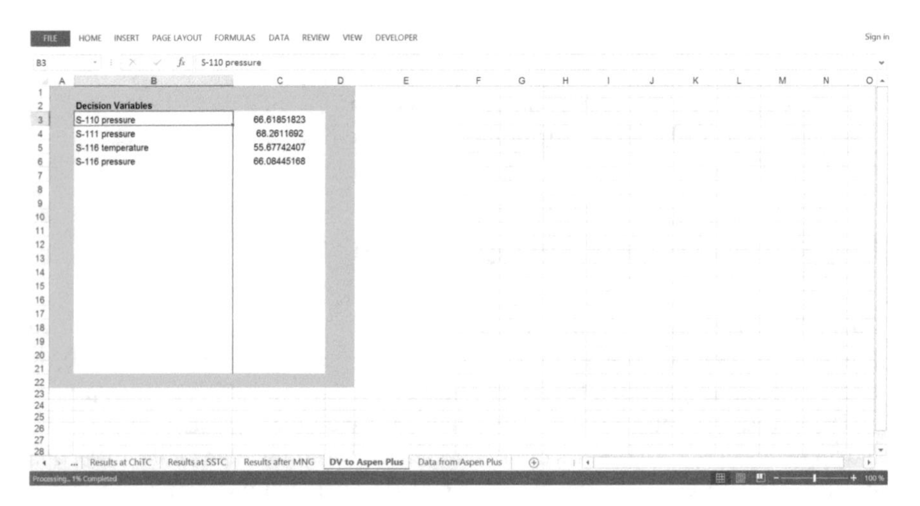

Fig. 7.4 MS Excel® sheet where decision variable values will be sent to the process simulator (case 2)

7.4 Link Between the Process Simulator and Optimization Algorithm

For the adequate link between the process simulator software (Aspen Plus® for this example) and the stochastic optimization algorithm (the I-MODE in this case), it is necessary to follow the methodology mentioned in previous chapters. It is recommendable to add two more MS Excel® sheets, the first one for the decision variable values that will be sent to the simulator (Fig. 7.4) and the second one for the response variable values that will be received from the simulator (Fig. 7.5).

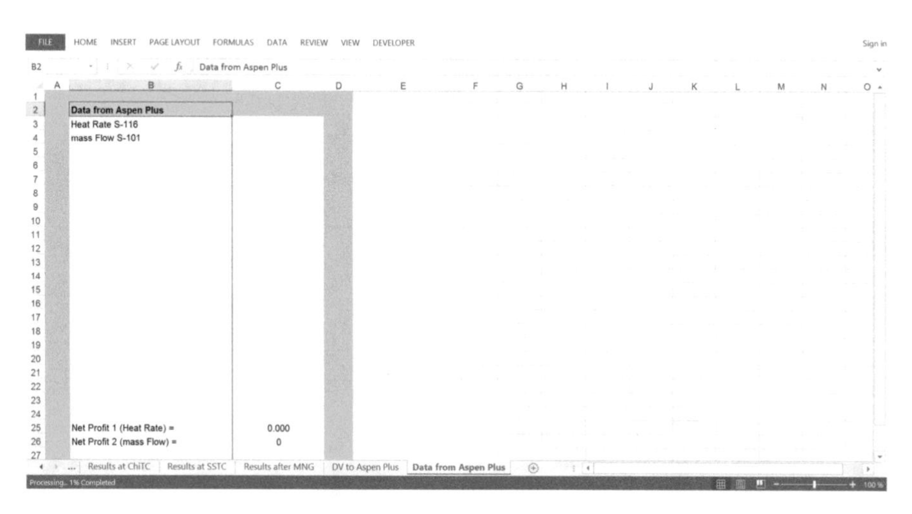

Fig. 7.5 MS Excel® sheet where response variable values will be received from the simulator (case 2)

7.5 Exercises

To download the example of the Generation Power Plant, please click on the following link:

http://extras.springer.com

1. Use the process flowsheet for the biodiesel production from degummed soybean oil implemented in SuperPro Designer® as shown in Fig. 7.2 and implement the following:

   (a) Implement a link between the SuperPro Designer® and the I-MODE algorithm in MS Excel®.

   (b) Analyze the Pareto curves obtained after running the optimization with the same selected decision variables for the lower and upper bounds shown in this chapter.

Appendix

In this part of the text, you can find some details of the topics treated in this book.

Appendix A: Code for the link between Aspen Plus and MS Excel®

```
  Public Sub REGENERATIVE_RANKINE_CYCLE()


  Dim Power_Plant As IHapp
 Set Power_Plant = GetObject("SIMULATOR FILE ROUTE\FILE NAME.bkp")
  Power_Plant.Visible = True
  Power_Plant.SuppressDialogs = True

  'DECISION VARIABLES TRANSFER FROM EXCEL TO ASPEN PLUS


  Dim Boiler_Temp As IHNode, Boiler_Pres As IHNode, TurbHP_Pres As
IHNode, TurbIP_Pres As IHNode, TurbLP_Pres As IHNode, Pump_Pres As
IHNode, Div1_Frac As IHNode, Div2_Frac As IHNode


    Set  Boiler_Temp  =  Power_Plant.Tree.FindNode("\Data\Blocks\
BOILER\Input\TEMP")
    Set  Boiler_Pres  =  Power_Plant.Tree.FindNode("\Data\Blocks\
BOILER\Input\PRES")
    Set  TurbHP_Pres  =  Power_Plant.Tree.FindNode("\Data\Blocks\
TURB-HP\Input\DELP")
```


J. M. Ponce-Ortega, L. G. Hernández-Pérez, *Optimization of Process Flowsheets through Metaheuristic Techniques*,
https://doi.org/10.1007/978-3-319-91722-1

```
    Set  TurbIP_Pres  =  Power_Plant.Tree.FindNode("\Data\Blocks\
TURB-IP\Input\DELP")
    Set  TurbLP_Pres  =  Power_Plant.Tree.FindNode("\Data\Blocks\
TURB-LP\Input\DELP")
    Set Pump_Pres = Power_Plant.Tree.FindNode("\Data\Blocks\PUMP\
Input\PRES")
   Set Div1_Frac = Power_Plant.Tree.FindNode("\Data\Blocks\SPLIT1\
Input\FRAC\3")
   Set Div2_Frac = Power_Plant.Tree.FindNode("\Data\Blocks\SPLIT2\
Input\FRAC\6")

  Boiler_Temp.Value = Sheets("DV to Aspen Plus").Range("C3")
  Boiler_Pres.Value = Sheets("DV to Aspen Plus").Range("C4")
  TurbHP_Pres.Value = Sheets("DV to Aspen Plus").Range("C5")
  TurbIP_Pres.Value = Sheets("DV to Aspen Plus").Range("C6")
  TurbLP_Pres.Value = Sheets("DV to Aspen Plus").Range("C7")
  Pump_Pres.Value = Sheets("DV to Aspen Plus").Range("C8")
  Div1_Frac.Value = Sheets("DV to Aspen Plus").Range("C9")
  Div2_Frac.Value = Sheets("DV to Aspen Plus").Range("C10")

  Power_Plant.Reinit     'REINITIALIZE ASPEN PLUS
  Power_Plant.Run        'ACTIVATE ASPEN PLUS

  'DATA TRANSFER FROM ASPEN HYSYS TO EXCEL


  Dim Boiler_Q As IHNode, TurbHP_W As IHNode, TurbIP_W As IHNode,
TurbLP_W As IHNode, Tower_Q As IHNode, Pump_W As IHNode, Boiler_
Cost As IHNode, Tower_Cost As IHNode, Boiler_Op_Cost As IHNode


   Set Boiler_Q = Power_Plant.Tree.FindNode("\Data\Blocks\BOILER\
Output\QNET")
  Set TurbHP_W = Power_Plant.Tree.FindNode("\Data\Blocks\TURB-HP\
Output\WNET")
  Set TurbIP_W = Power_Plant.Tree.FindNode("\Data\Blocks\TURB-IP\
Output\WNET")
  Set TurbLP_W = Power_Plant.Tree.FindNode("\Data\Blocks\TURB-LP\
Output\WNET")
   Set  Tower_Q  =  Power_Plant.Tree.FindNode("\Data\Blocks\TOWER\
Output\HX_DUTY")
    Set  Pump_W  =  Power_Plant.Tree.FindNode("\Data\Blocks\PUMP\
Output\WNET")
    Set  Boiler_Cost  =  Power_Plant.Tree.FindNode("\Data\Blocks\
BOILER\Output\UTIL_COST")
```

```
  Set Tower_Cost = Power_Plant.Tree.FindNode("\Data\Blocks\TOWER\
Output\UTL_COST")

  Sheets("Data from Aspen Plus").Range("C3") = Boiler_Q.Value
  Sheets("Data from Aspen Plus").Range("C4") = TurbHP_W.Value
  Sheets("Data from Aspen Plus").Range("C5") = TurbIP_W.Value
  Sheets("Data from Aspen Plus").Range("C6") = TurbLP_W.Value
  Sheets("Data from Aspen Plus").Range("C7") = Tower_Q.Value
  Sheets("Data from Aspen Plus").Range("C8") = Pump_W.Value
  Sheets("Data from Aspen Plus").Range("C9") = Boiler_Cost.Value
  Sheets("Data from Aspen Plus").Range("C10") = Tower_Cost.Value

  End Sub
```

Appendix B: Code for the link between SuperPro Designer® and MS Excel®

```
  Public Sub SetAdReadStarchFlowrate()
  Dim str1, str2, str3, str4 As String
  Dim var1, var2, var3, var4, var5, var6 As Variant
  Set DocObj1 = GetObject("SIMULATOR FILE ROUTE\FILE NAME.spf")
  var1 = Worksheets("DV to Aspen Plus").Range("C3")
  var2 = Worksheets("DV to Aspen Plus").Range("C4")
  var3 = Worksheets("DV to Aspen Plus").Range("C5")
  var4 = Worksheets("DV to Aspen Plus").Range("C6")
  str1 = "S-101"
  str2 = "S-102"
  str3 = "S-110"
  str4 = "S-115"
  DocObj1.SetStreamVarVal str1, VarID.temperature_VID, var1, str1
  DocObj1.SetStreamVarVal str1, VarID.pressure_VID, var2, str1
  DocObj1.SetStreamVarVal str2, VarID.temperature_VID, var3, str2
  DocObj1.SetStreamVarVal str2, VarID.pressure_VID, var4, str2
  DocObj1.DoMEBalances var1, var2, var3, var4, var5, var6
  DocObj1.DoEconomicCalculations
  DocObj1.GetStreamVarVal str3, VarID.HeatRate_VID, var5, str3
  DocObj1.GetStreamVarVal str4, VarID.massFlow_VID, var6, str4
  Worksheets("Data from Aspen Plus").Range("C3") = var5
  Worksheets("Data from Aspen Plus").Range("C4") = var6

  End Sub
```

Appendix C: Code for the link between MS Excel® and MATLAB®

File 1: Function name and linker program file route

```
function F=g(pop)
pop

excelObject=actxserver('Excel.Application');
excelObject.Workbooks.Open ('LIKER FILE ROUTE\FILE NAME.xlsm')
hworkbook=excelObject.workbooks;
F=excelObject.Run ('Rankine',pop(1),pop(2))
hworkbook.Close;
F=-F
```

File 2: Optimization parameters and function call

```
clc;
clear;
lb=[100 5]';
ub=[500 100]';
opts = gaoptimset('Generations',10,'PopInitRange',[0 0;500
100],'PopulationSize',10,'EliteCount',2,'CrossoverFraction',0.8,'
TimeLimit',Inf,'StallGenLimit',Inf,'StallTimeLimit',inf,'Display'
,'iter','PlotFcns',@gaplotbestf);
[x,fval,reason,output] = ga(@g,2,[],[],[],[],lb,ub,[],opts)
```

Bibliography

AspenTech. (2009). *Aspen Process Evaluator User's Guide, V7.1*. s.l.: Aspen Technology Inc.

AspenTech. (2015). *Aspen Plus User's Guide, V8.8*. s.l.: Aspen Technology Inc.

Bamufleh, H. S., Ponce-Ortega, J. M., & El-Halwagi, M. M. (2013). Multi-objective optimization of process cogeneration systems with economic, environmental, and social tradeoffs. *Clean Technologies and Environmental Policy, 15*(1), 185–197.

Birnbaum, D. (2005). *Excel VBA programming for the absolute beginner*. Boston, MA: Thomson Course Technology PTR.

Coello-Coello, C. A., Van-Veldhuizen, D. A., & Lamont, G. B. (2002). *Evolutionary algorithms for solving multi-objective problems*. New York, NY: Kluwer Academic.

Contreras-Zarazúa, G., et al. (2016). Multi-objective optimization involving cost and control properties in reactive distillation processes to produce diphenyl carbonate. *Computers & Chemical Engineering, 105*, 185–196.

Devillers, J. (1996). *Genetic algorithms in molecular modeling*. San Diego, CA: Academic Press.

Feng, Y., & Rangaiah, G. P. (2011). *Evaluating capital cost estimation programs*. Singapore: National University of Singapore.

Gen, M., & Cheng, R. (1997). *Genetic algorithms and engineering design*. Ashikaga: Wiley.

Goldberg, D. E. (1989). Genetic algorithms and Walsh functions: Part I, A gentle introduction. *Complex Systems, 3*, 129–152.

Gómez-Ríos, D., Barrera-Zapata, R., & Ríos-Estepa, R. (2017). Comparison of process technologies for chitosan production from shrimp shell waste: A techno-economic approach using Aspen Plus. *Food and Bioproducts Processing, 103*, 49–57.

González-Bravo, R., Nápoles-Rivera, F., Ponce-Ortega, J. M., & El-Halwagi, M. M. (2016). Multiobjective optimization of dual-purpose power plants and water distribution networks. *ACS Sustainable Chemistry & Engineering, 4*(12), 6852–6866.

Guo, H., et al. (2014). Differential evolution improved with self-adaptive control parameters based on simulated annealing. *Swarm and Evolutionary Computation, 19*, 52–67.

Gutiérrez-Arriaga, C. G., Serna-González, M., Ponce-Ortega, J. M., & El-Halwagi, M. M. (2013). Multi-objective optimization of steam power plants for sustainable generation of electricity. *Clean Technologies and Environmental Policy, 15*, 551–556.

Gutiérrez-Arriaga, C. G., et al. (2015). Industrial waste heat recovery and cogeneration involving organic Rankine cycles. *Clean Technologies and Environmental Policy, 17*(3), 767–779.

Haas, M. J., McAloon, A. J., & Ye, W. C. (2006). A process model to estimate biodiesel production costs. *Bioresource Technology, 97*, 671–678.

J. M. Ponce-Ortega, L. G. Hernández-Pérez, *Optimization of Process Flowsheets through Metaheuristic Techniques*,
https://doi.org/10.1007/978-3-319-91722-1

Hauck, M., Herrmann, S., & Spliethoff, H. (2017). Simulation of a reversible SOFC with Aspen Plus. *International Journal of Hydrogen Energy, 42*(15), 10329–10340.

Hock, W., & Schittkowski, K. (1981). *Test examples for nonlinear programming codes*. New York, NY: Springer-Verlag.

Huang, Y. Y. (2010). *Analysis of capital cost estimation methods and computer programs*. Singapore: Department of Chemical & Biomolecular Engineering, National University of Singapore.

Kirkpatrick, S., Gelatt, C. D., & Vecchi, M. P. (1983). Optimization by Simulated Annealing. *Science, 220*(4598), 671–680.

Martinez-Gomez, J., Nápoles-Rivera, F., Ponce-Ortega, J. M., & El-Halwagi, M. M. (2017). Optimization of the production of syngas from shale gas with economic and safety considerations. *Applied Thermal Engineering, 110*, 678–685.

Medina-Herrera, N., Tututi-Avila, S., Jiménez-Gutiérrez, A., & Segovia-Hernández, J. G. (2017). Optimal design of a multi-product reactive distillation system for silanes production. *Computers & Chemical Engineering, 105*, 132–141.

Morgan, J. C., et al. (2017). Thermodynamic modeling and uncertainty quantification of CO_2-loaded aqueous MEA solutions. *Chemical Engineering Science, 168*, 309–324.

Ouyang, A., et al. (2015). A novel hybrid multi-objective population migration algorithm. *International Journal of Pattern Recognition and Artificial Intelligence, 29*(01), 1559001.

Pauls, J. H., Mahinpey, N., & Mostafavi, E. (2016). Simulation of air-steam gasification of woody biomass in a bubbling fluidized bed using Aspen Plus: A comprehensive model including pyrolysis, hydrodynamics and tar production. *Biomass and Bioenergy, 95*, 157–166.

Ponce-Ortega, J. M., Nápoles-Rivera, F., El-Halwagi, M. M., & Jiménez-Gutiérrez, A. (2012). An optimization approach for the synthesis of recycle and reuse water integration networks. *Clean Technologies and Environmental Policy, 14*(1), 133–151.

Ponce-Ortega, J. M., Serna-González, M., & Jiménez-Gutiérrez, A. (2009). Use of genetic algorithms for the optimal design of shell-and-tube heat exchangers. *Applied Thermal Engineering, 29*(2–3), 203–209.

Quiroz-Ramírez, J. J., et al. (2017a). Multiobjective stochastic optimization approach applied to a hybrid process production–separation in the production of biobutanol. *Industrial & Engineering Chemistry Research, 56*(7), 1823–1833.

Quiroz-Ramírez, J. J., et al. (2017b). Optimal selection of feedstock for biobutanol production considering economic and environmental aspects. *ACS Sustainable Chemistry & Engineering, 5*, 4018–4030.

Sandler, S. I. (2015). *Using Aspen Plus in thermodynamics instruction* (1st ed.). Hoboken, NJ: Wiley.

Santibañez-Aguilar, J. E., Morales-Rodriguez, R., González-Campos, J. B., & Ponce-Ortega, J. M. (2016). Stochastic design of biorefinery supply chains considering economic and environmental objectives. *Journal of Cleaner Production, 136*, 224–245.

Segovia-Hernández, J. G., & Gómez-Castro, F. I. (2017). *Stochastic process optimization using Aspen Plus®*. s.l.: s.n.

Seider, W. D., Seader, J. D., & Lewin, D. R. (2010). *Product and process design principles* (3rd ed.). New York, NY: Wiley.

Sharma, S., Rangaiah, G. P. & Cheah, K. S. (2012). Multi-objective optimization using MS Excel with an application to design of a falling-film evaporator system. Food and Bioproducts Processing, 90(2), 123–134.

Sharma, S., & Rangaiah, G. P. (2013). An improved multi-objective differential evolution with a termination criterion for optimizing chemical processes. *Computers & Chemical Engineering, 56*, 155–173.

Sharma, S., & Rangaiah, G. P. (2014). Hybrid approach for multiobjective optimization and its application to process engineering problems. In J. Valadi & P. Siarry (Eds.), *Applications of metaheuristics in process engineering* (pp. 423–444). s.l.: Springer International Publishing.

Sharma, S., & Rangaiah, G. P. (2016). Mathematical modeling simulation and optimization for process design. In G. P. Rangaiah (Ed.), *Chemical process retrofitting and revamping: Techniques and applications* (pp. 99–128). Singapore: Wiley.

Sinnott, R. K., & Towler, G. (2009). *Chemical engineering design* (5th ed.). Burlington, MA: Butterworth and Heinemann.

Turton, R., Bailie, R. C., Whiting, W. B., & Shaeiwitz, J. A. (2009). *Analysis, synthesis, and design of chemical processes* (3rd ed.). Upper Saddle River, NJ: Prentice Hall.

Ulrich, G. D., & Vasudevan, P. T. (2004). *Chemical engineering process design and economics – a practical guide* (2nd ed.). Durham, NH: Process Publishing.

Vasudevan, P. T., & Agrawal, D. A. (1999). Software package for capital cost estimation. *Chemical Engineering Education, 33*(3), 254–256.

Wang, X., & Tang, L. (2013). Multiobjective operation optimization of naphtha pyrolysis process using parallel differential evolution. *Industrial & Engineering Chemistry Research, 52*(40), 14415–14428.

Woinaroschy, A. (2009). Simulation and optimization of citric acid production with SuperPro designer using a client-server interface. *CHIM, 9*, 979–983.

Wong, C. L. (2010). *Development and application of a capital cost estimation program*. Singapore: Department of Chemical & Biomolecular Engineering, National University of Singapore.

Wong, J. Y., Sharma, S., & Rangaiah, G. P. (2016). Design of shell-and-tube heat exchangers for multiple objectives using elitist non-dominated sorting genetic algorithm with termination criteria. *Applied Thermal Engineering, 93*, 888–899.

Websites

aspentech.com
honeywellprocess.com
intelligent.com
mathworks.com
microsoft.com
psenterprise.com
software.schneider-electric.com

Index

J. M. Ponce-Ortega, L. G. Hernández-Pérez, *Optimization of Process Flowsheets through Metaheuristic Techniques*,
https://doi.org/10.1007/978-3-319-91722-1

Printed in the United States
By Bookmasters